M208 Pur

The Open University

GTA4

Cosets and Lagrange's Theorem

This publication forms part of an Open University course. Details of this and other Open University courses can be obtained from the Student Registration and Enquiry Service, The Open University, PO Box 197, Milton Keynes, MK7 6BJ, United Kingdom: tel. +44 (0)870 333 4340, e-mail general-enquiries@open.ac.uk

Alternatively, you may visit the Open University website at http://www.open.ac.uk where you can learn more about the wide range of courses and packs offered at all levels by The Open University.

To purchase a selection of Open University course materials, visit the webshop at www.ouw.co.uk, or contact Open University Worldwide, Michael Young Building, Walton Hall, Milton Keynes, MK7 6AA, United Kingdom, for a brochure: tel. +44 (0)1908 858785, fax +44 (0)1908 858787, e-mail ouwenq@open.ac.uk

The Open University, Walton Hall, Milton Keynes, MK7 6AA.

First published 2006.

Edited, designed and typeset by The Open University, using the Open University TeX System.

Printed and bound in the United Kingdom by Hobbs the Printers Limited, Brunel Road, Totton, Hampshire SO40 3WX.

ISBN 0 7492 0218 1

1.1

Contents

Introduction

This unit is concerned with the structure of groups. We present concepts and strategies which help us to recognise some important properties of groups and their subgroups.

In Section 1 we introduce the idea of a *coset* and look at many examples. We investigate properties of cosets; in particular, we find that cosets *partition* a group.

In Section 2 we turn to *Lagrange's Theorem*, a powerful result that relates the order of a group to the orders of its subgroups. The proof of this theorem is straightforward, following from properties of cosets established in the previous section. The importance of the theorem begins to emerge as we discover simple but important consequences of the result.

In Section 3 we use Lagrange's Theorem and other results of Section 2 to determine all the possible structures—all the isomorphism classes—of groups of order 1 to 7, and we describe the isomorphism classes of groups of order 8.

In Section 4 we compare the *left* and *right* cosets of subgroups. This leads us to the notion of a *normal subgroup*. This is a crucial concept: normal subgroups are the building blocks of group theory.

In Sections 5 we introduce the *quotient structure*. We find that the set of cosets of a normal subgroup forms a new group with a binary operation inherited from the binary operation of the original group. Such a group is called a *quotient group*. We look at examples of quotient groups of finite groups, and we see how to construct the group table for such a quotient group.

In Section 6 we investigate quotient groups of some infinite groups.

Study guide

The sections should be read in the natural order.

Section 1 sets the scene for the whole unit, by introducing the concept of a *coset*. It is essential to understand this concept before progressing, as the remainder of the unit depends on it. The video section is at the end of Section 1 and provides an overview of many of the key ideas in Sections 1–5. It can be watched at any time during your study of the unit, although you will find it easier to follow if you have first read, or at least skimmed through, Section 1.

It is important that you are able to identify the isomorphism class of a group of small order, as covered in Section 3.

Section 5 is the audio section.

Section 6 is not a long section, but you may find it difficult conceptually. The ideas introduced here are developed in Group Theory Block B, and you may find it helpful to reread this section after this further study of quotient groups.

1   Cosets

After working through this section, you should be able to:

(a) explain the meaning of the term *coset*;

(b) determine the cosets gH of a given subgroup H in a given group G;

(c) understand that cosets *partition a group* and determine such partitions in simple cases.

1.1   Introducing cosets

In this section we show how we can use a subgroup of a given finite group to split the group into smaller sets, each containing the same number of elements as the subgroup. We have already seen several instances of this.

For example, consider the following group $S(\square)$.

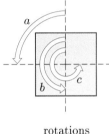
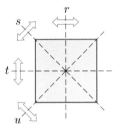

rotations   reflections

$\circ$	e	a	b	c	r	s	t	u
e	e	a	b	c	r	s	t	u
a	a	b	c	e	s	t	u	r
b	b	c	e	a	t	u	r	s
c	c	e	a	b	u	r	s	t
r	r	u	t	s	e	c	b	a
s	s	r	u	t	a	e	c	b
t	t	s	r	u	b	a	e	c
u	u	t	s	r	c	b	a	e

We can write the set of eight elements of $S(\square)$ as the union of two subsets of four elements:

$$S(\square) = \{e, a, b, c\} \cup \{r, s, t, u\}.$$

The subset $H = \{e, a, b, c\}$ is the subgroup of *direct* symmetries (rotations); the subset $\{r, s, t, u\}$ is the set of *indirect* symmetries (reflections). This way of splitting the set $S(\square)$ highlights one aspect of the structure of the group. Moreover, we can obtain each element of the second set by composing each element of H, the subgroup of direct symmetries, with a fixed indirect symmetry; for example, composing each direct symmetry with r on the left, we get

$$r \circ e = r, \quad r \circ a = u, \quad r \circ b = t, \quad r \circ c = s.$$

Next consider the group $\mathbb{Z}_{12}$. This can be split into four subsets by placing elements in the same subset if they have the same remainder when divided by 4:

$$\mathbb{Z}_{12} = \{0, 4, 8\} \cup \{1, 5, 9\} \cup \{2, 6, 10\} \cup \{3, 7, 11\}.$$

The subset $H = \{0, 4, 8\}$ is a subgroup of $\mathbb{Z}_{12}$, and the elements of successive subsets are obtained by adding an element which is not in H to each of the elements of H. For example, we can obtain the second subset by adding 1 (or 5 or 9) to each of the elements of H.

Each of these ways of splitting the groups $S(\square)$ and $\mathbb{Z}_{12}$ is a *partition* of the group into disjoint subsets, where one of the subsets is a *subgroup* of the group. These subsets are examples of *cosets*, which we now define formally.

The idea of a *partition* is introduced Unit I3, Section 4.1.

Definition   Let H be a subgroup of a group $(G, \circ)$ and let g be an element of G. The **coset** gH of H in G is the set of elements of G of the form $g \circ h$, where $h \in H$. That is,

$$gH = \{g \circ h : h \in H\},$$

which is the set obtained by composing each element of H with g on the left.

If H is finite, say

$$H = \{h_1, h_2, \ldots, h_m\},$$

then

$$gH = \{g \circ h_1, g \circ h_2, \ldots, g \circ h_m\}$$

Example 1.1   Determine all the distinct cosets of the subgroup $H = \{e, r\}$ in the group $S(\square)$.

Solution   The group $S(\square)$ has eight elements, so there are eight potential cosets—namely,

$$eH, \quad aH, \quad bH, \quad cH, \quad rH, \quad sH, \quad tH, \quad uH.$$

However, if we expect these to partition $S(\square)$, then we should expect to have only *four distinct* cosets.

Now

$$
\begin{aligned}
eH &= e\{e, r\} = \{e \circ e, e \circ r\} = \{e, r\}, \\
aH &= a\{e, r\} = \{a \circ e, a \circ r\} = \{a, s\}, \\
bH &= b\{e, r\} = \{b \circ e, b \circ r\} = \{b, t\}, \\
cH &= c\{e, r\} = \{c \circ e, c \circ r\} = \{c, u\}, \\
rH &= r\{e, r\} = \{r \circ e, r \circ r\} = \{r, e\}, \\
sH &= s\{e, r\} = \{s \circ e, s \circ r\} = \{s, a\}, \\
tH &= t\{e, r\} = \{t \circ e, t \circ r\} = \{t, b\}, \\
uH &= u\{e, r\} = \{u \circ e, u \circ r\} = \{u, c\}.
\end{aligned}
$$

So there are only four distinct cosets of H in $S(\square)$:

$$
\begin{aligned}
eH &= rH = \{e, r\}, \\
aH &= sH = \{a, s\}, \\
bH &= tH = \{b, t\}, \\
cH &= uH = \{c, u\}. \quad \blacksquare
\end{aligned}
$$

Example 1.2   Determine all the distinct cosets of the subgroup $H = \{e, (1\ 2\ 3), (1\ 3\ 2)\}$ in S_3.

Here H is the alternating group A_3 introduced in Unit GTA3, Section 2.3.

Solution   The group S_3 has six elements, so there are six potential cosets—namely,

$$eH, \quad (1\ 2)H, \quad (1\ 3)H, \quad (2\ 3)H, \quad (1\ 2\ 3)H, \quad (1\ 3\ 2)H,$$

but if we expect these to partition S_3, then we should expect to have only two distinct cosets.

Now

$$eH = \{e \circ e, e \circ (1\ 2\ 3), e \circ (1\ 3\ 2)\}$$
$$= \{e, (1\ 2\ 3), (1\ 3\ 2)\} = H,$$
$$(1\ 2)H = \{(1\ 2) \circ e, (1\ 2) \circ (1\ 2\ 3), (1\ 2) \circ (1\ 3\ 2)\}$$
$$= \{(1\ 2), (2\ 3), (1\ 3)\},$$
$$(1\ 3)H = \{(1\ 3) \circ e, (1\ 3) \circ (1\ 2\ 3), (1\ 3) \circ (1\ 3\ 2)\}$$
$$= \{(1\ 3), (1\ 2), (2\ 3)\},$$
$$(2\ 3)H = \{2\ 3) \circ e, (2\ 3) \circ (1\ 2\ 3), (2\ 3) \circ (1\ 3\ 2)\}$$
$$= \{(2\ 3), (1\ 3), (1\ 2)\},$$
$$(1\ 2\ 3)H = \{(1\ 2\ 3) \circ e, (1\ 2\ 3) \circ (1\ 2\ 3), (1\ 2\ 3) \circ (1\ 3\ 2)\}$$
$$= \{(1\ 2\ 3), (1\ 3\ 2), e\} = H,$$
$$(1\ 3\ 2)H = \{(1\ 3\ 2) \circ e, (1\ 3\ 2) \circ (1\ 2\ 3), (1\ 3\ 2) \circ (1\ 3\ 2)\}$$
$$= \{(1\ 3\ 2), e, (1\ 2\ 3)\} = H.$$

$\circ$	e	(123)	(132)	(23)	(13)	(12)
e	e	(123)	(132)	(23)	(13)	(12)
(123)	(123)	(132)	e	(12)	(23)	(13)
(132)	(132)	e	(123)	(13)	(12)	(23)
(23)	(23)	(13)	(12)	e	(123)	(132)
(13)	(13)	(12)	(23)	(132)	e	(123)
(12)	(12)	(23)	(13)	(123)	(132)	e

$$S_3$$

So there are only two distinct cosets of H in S_3:

$$eH = (1\ 2\ 3)H = (1\ 3\ 2)H = H,$$
$$(1\ 2)H = (1\ 3)H = (2\ 3)H = \{(1\ 2), (1\ 3), (2\ 3)\}. \quad \blacksquare$$

Exercise 1.1 For each of the following subgroups H of $S(\square)$, determine all eight potential cosets—namely,

$$eH, \quad aH, \quad bH, \quad cH, \quad rH, \quad sH, \quad tH, \quad uH.$$

How many distinct cosets are there in each case?

(a) $H = \{e, s\}$ (b) $H = \{e, b, r, t\}$

Exercise 1.2 Determine all the distinct cosets of the subgroup $H = \{e, (1\ 2)\}$ in S_3.

In studying the examples and working through these exercises, you may have noticed the following common features.

Properties of cosets

1. For each element g and each subgroup H of a *finite* group, the coset gH has the same number of elements as H.

2. For each element g and each subgroup H, the element g lies in the coset gH.

3. One of the cosets gH is H itself.

4. Any two cosets $g_1 H$ and $g_2 H$ are either the same set or are disjoint.

For example, we have seen the following properties for cosets of the subgroup $H = \{e, r\}$ in $S(\square)$:

each coset has two elements;

$a = a \circ e$, so $a \in aH$, and similarly for the other elements;

$cH = rH = H$;

aH and sH are the same, and aH and rH are disjoint.

Remarks

1. We can show that property 1 is true in general, as follows.

If $H = \{h_1, h_2, \ldots, h_m\}$ is a subgroup of order m, then the coset

$$gH = \{g \circ h_1, g \circ h_2, \ldots, g \circ h_m\}$$

has at most m elements. The m listed elements in gH must be distinct because the Cancellation Laws tell us that $g \circ h_i = g \circ h_j$ only when $h_i = h_j$, so gH has *exactly* m elements.

The Cancellation Laws are given in Unit GTA1, Property 4.7.

2. Property 2 holds because the identity element e belongs to the subgroup H, so $g = g \circ e$ lies in the coset gH.

3. The subgroup H is a coset because

    $$eH = \{e \circ h : h \in H\} = \{h : h \in H\} = H.$$

    In fact, if g is any element of H, then $gH = H$ (see Exercise 1.3 below).

4. Property 4 means that two cosets are either the same or have no element in common.

We prove property 4 in Section 1.2.

Exercise 1.3 Let H be a subgroup of a finite group G and let $g \in G$. Show that

$$\text{if } g \in H, \quad \text{then} \quad gH = H,$$

and, conversely,

$$\text{if } gH = H, \quad \text{then} \quad g \in H.$$

This result also holds for infinite groups, but we do not prove it here.

1.2 Partitioning a group into cosets

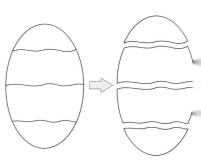

We now focus on the main property of cosets which has been illustrated by all our examples—the fact that cosets partition a group. Recall that a *partition* of a set splits the set into (disjoint) subsets such that each element of the set belongs to one and only one subset.

We have seen that the group $S(\square)$ can be partitioned into four cosets, each with two elements:

$$\{e, r\}, \quad \{a, s\}, \quad \{b, t\}, \quad \{c, u\}.$$

Example 1.1

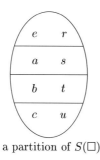

a partition of $S(\square)$

We have also seen that the group S_3 can be partitioned into three cosets, each with two elements:

$$\{e, (1\ 2)\}, \quad \{(1\ 3), (1\ 2\ 3)\}, \quad \{(2\ 3), (1\ 3\ 2)\};$$

Exercise 1.2

and into two cosets, each with three elements:

$$\{e, (1\ 2\ 3), (1\ 3\ 2)\}, \quad \{(1\ 2), (1\ 3), (2\ 3)\}.$$

Example 1.2

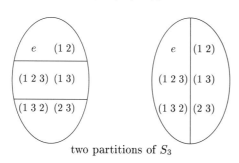

two partitions of S_3

We shall prove that this happens in general: in *any* group the cosets of *any* subgroup *partition* the group. This means that each element of the group belongs to one and only one of the distinct cosets.

To prove this result, we use equivalence relations. We know that the equivalence classes of an equivalence relation form a partition of the set on which the relation is defined, so we define an equivalence relation on the set of elements of a group in such a way that the equivalence classes are the cosets of the given subgroup.

Equivalence relations are discussed in Unit I3, Section 4.

Theorem 1.1   Let H be a subgroup of a group G. Then the cosets of H form a partition of G.

Comments

This definition means that

$x \sim y$ if x belongs to the coset yH.

We show that $\sim$ is reflexive, symmetric and transitive.

We need to show that $x \sim x$.

Here we use property 2 for cosets.

We need to show that

if $x \sim y$, then $y \sim x$.

We assume that $x \sim y$, and write down what this means;

then we translate our statement about cosets into a statement about elements.

H is a *subgroup* of G.

We have x in terms of y; we want to express y in terms of x.

We use the group axioms to do the necessary algebra.

We have expressed y as a product of x and an element in H, so $y \in xH$.

We need to show that

if $x \sim y$ and $y \sim z$, then $x \sim z$.

We assume that $x \sim y$ and $y \sim z$, and write down what this means;

then we translate our statements about cosets into statements about elements.

Proof

We define a relation $\sim$ on G by

$x \sim y$ if and only if $x \in yH$.

We show that $\sim$ is an equivalence relation.

E1 REFLEXIVE

For each element x in the group G,

$$x \in xH,$$

so $x \sim x$, and the relation is reflexive.

E2 SYMMETRIC

Suppose that $x \sim y$; then

$$x \in yH,$$

so

$$x = yh, \text{ for some element } h \text{ in } H.$$

Now h has an inverse h^{-1} in H.

Composing both sides of the above equation with h^{-1} on the right, we obtain

$$\begin{aligned}
xh^{-1} &= (yh)h^{-1} \\
&= y(hh^{-1}) \quad \text{(associativity)} \\
&= ye \quad \text{(inverses)} \\
&= y \quad \text{(identity)}.
\end{aligned}$$

Thus $y = xh^{-1}$, so $y \in xH$;
so $y \sim x$, and the relation is symmetric.

E3 TRANSITIVE

Suppose that $x \sim y$ and $y \sim z$; then

$$x \in yH \text{ and } y \in zH,$$

so

$$x = yh_1 \text{ and } y = zh_2,$$

for some elements h_1, h_2 in H.

We have x in terms of y, and y in terms of z; we want x in terms of z.

Substituting the expression for y in the expression for x, we obtain

$$\begin{aligned} x &= (zh_2)h_1 \\ &= z(h_2 h_1) \quad \text{(associativity)} \\ &= zh_3 \quad \text{(closure of } H) \end{aligned}$$

We use the group and subgroup axioms.

We have expressed x as a product of z and an element in H, so $x \in zH$.

for some element $h_3 \in H$, so $x \in zH$, and the relation is transitive.

Hence $\sim$ is an equivalence relation because it satisfies the axioms E1, E2 and E3.

Thus $\sim$ partitions G into equivalence classes.

Now

The equivalence class $[\![x]\!]$ of x is the set of all elements $y \in G$ which are related to x under $\sim$; this is just the coset xH.

$$\begin{aligned} [\![x]\!] &= \{y \in G : y \sim x\} \\ &= \{y \in G : y \in xH\} \\ &= xH, \end{aligned}$$

so the equivalence classes are the cosets of H.

This completes the proof. ∎

Exercise 1.4 Using the properties of cosets on page 7, determine the partition of $S(\square)$ into cosets of each of the following subgroups H.

(a) $H = \{e, a, b, c\}$

(b) $H = \{e\}$

(c) $H = S(\square)$

Remark Property 4 follows from Theorem 1.1, since it implies that two cosets $g_1 H$ and $g_2 H$ are either the same or have no elements in common. Thus we may use property 4 to note that

if $g_2 \in g_1 H$,   then $g_2 H = g_1 H$;

if $g_2 H = g_1 H$,   then $g_2 \in g_1 H$ and $g_1 \in g_2 H$.

Before looking at further examples of cosets and partitions, we describe a strategy for partitioning a given finite group into cosets; it is based on the above observations.

Strategy 1.1 To partition a finite group G into cosets of a given subgroup H.

1. Take H as the first coset.

2. Choose any element g not yet assigned to a coset and determine the coset gH to which g belongs.

3. Repeat step 2 until every element of G has been assigned to a coset.

Property 1 states that all cosets of H have the same number of elements, so if $|G| = n$ and $|H| = m$, this process will end after we have found n/m cosets.

Example 1.3   Partition the group $S(\square)$ into cosets of the subgroup $H = \{e, b\}$.

Solution   We use Strategy 1.1.

1. We take the subgroup $\{e, b\}$ as the first coset.

2. We choose any element not in $\{e, b\}$, say a.

    We determine aH, the second coset:

    $$aH = a\{e, b\} = \{a \circ e, a \circ b\} = \{a, c\}.$$

    We choose any element not in H or aH, say r.

    We determine rH, the third coset:

    $$rH = r\{e, b\} = \{r \circ e, r \circ b\} = \{r, t\}.$$

    We choose any element not in H, aH or rH, say s.

    We determine sH, the fourth coset:

    $$sH = s\{e, b\} = \{s \circ e, s \circ b\} = \{s, u\}.$$

3. Every element of $S(\square)$ has now been assigned to a coset, so the partition is

    $$\{e, b\}, \quad \{a, c\}, \quad \{r, t\}, \quad \{s, u\}. \quad \blacksquare$$

Our next example also follows Strategy 1.1.

Example 1.4   Partition the alternating group A_4 into cosets of the subgroup

See Unit GTA3, Section 2.3.

$$H = \{e, (1\ 2)(3\ 4), (1\ 3)(2\ 4), (1\ 4)(2\ 3)\}.$$

Solution   The group A_4 consists of the twelve even permutations in S_4. There are four elements in the subgroup H, and the remaining eight elements are the 3-cycles:

$$(1\ 2\ 3), \quad (1\ 3\ 2), \quad (1\ 2\ 4), \quad (1\ 4\ 2),$$
$$(1\ 3\ 4), \quad (1\ 4\ 3), \quad (2\ 3\ 4), \quad (2\ 4\ 3).$$

1. We take the subgroup H as the first coset:

    $$\{e, (1\ 2)(3\ 4), (1\ 3)(2\ 4), (1\ 4)(2\ 3)\}.$$

2. To find another coset, we choose any even permutation not in H, say $(1\ 2\ 3)$.

    We determine $(1\ 2\ 3)H$, the second coset:

    $$\begin{aligned}
    (1\ 2\ 3)H &= \{(1\ 2\ 3) \circ e, (1\ 2\ 3) \circ (1\ 2)(3\ 4),\\
    &\quad (1\ 2\ 3) \circ (1\ 3)(2\ 4), (1\ 2\ 3) \circ (1\ 4)(2\ 3)\}\\
    &= \{(1\ 2\ 3), (1\ 3\ 4), (2\ 4\ 3), (1\ 4\ 2)\}.
    \end{aligned}$$

    We choose an element not in H or $(1\ 2\ 3)H$, say $(1\ 3\ 2)$.

    We determine $(1\ 3\ 2)H$, the third coset:

    $$\begin{aligned}
    (1\ 3\ 2)H &= \{(1\ 3\ 2) \circ e, (1\ 3\ 2) \circ (1\ 2)(3\ 4),\\
    &\quad (1\ 3\ 2) \circ (1\ 3)(2\ 4), (1\ 3\ 2) \circ (1\ 4)(2\ 3)\}\\
    &= \{(1\ 3\ 2), (2\ 3\ 4), (1\ 2\ 4), (1\ 4\ 3)\}.
    \end{aligned}$$

3. Every element of A_4 has been assigned to a coset, so the partition is

$$\{e, (1\ 2)(3\ 4), (1\ 3)(2\ 4), (1\ 4)(2\ 3)\},$$
$$\{(1\ 2\ 3), (1\ 3\ 4), (2\ 4\ 3), (1\ 4\ 2)\},$$
$$\{(1\ 3\ 2), (2\ 3\ 4), (1\ 2\ 4), (1\ 4\ 3)\}.$$

The twelve elements of A_4 are now divided into three cosets, each with four elements.   ■

Exercise 1.5   Partition S_3 into cosets of the subgroup $H = \{e, (1\ 3)\}$.

1.3 Cosets in additive groups

We now turn to groups where the binary operation is *additive*—that is, addition or modular addition. For additive groups, we write a coset as $g + H$, rather than gH.

This is just a reminder that the group operation is addition.

Convention   Let $(G, +)$ be a group and H be a subgroup of G. The coset of H containing the element $g \in G$ is the set

$$g + H = \{g + h : h \in H\}.$$

Example 1.5   Determine the partition of the group $\mathbb{Z}_9$ into cosets of the subgroup $H = \{0, 3, 6\}$.

Solution   We use Strategy 1.1 for partitioning a group into cosets.

1. We take the subgroup H as the first coset.

2. To find another coset, we choose any element of $\mathbb{Z}_9$ which is not in H, say 1.

    We determine $1 + H$, the second coset:

    $$\begin{aligned}1 + H &= 1 + \{0, 3, 6\}\\ &= \{1 +_9 0,\ 1 +_9 3,\ 1 +_9 6\} = \{1, 4, 7\}.\end{aligned}$$

    To find a third coset, we choose an element which does not belong to either of the cosets found so far, say 8.

    We determine $8 + H$, the third coset:

    $$\begin{aligned}8 + H &= 8 + \{0, 3, 6\}\\ &= \{8 +_9 0,\ 8 +_9 3,\ 8 +_9 6\} = \{8, 2, 5\}.\end{aligned}$$

3. Every element of $\mathbb{Z}_9$ has been assigned to a coset, so the partition is

    $$\{0, 3, 6\}, \quad \{1, 4, 7\}, \quad \{2, 5, 8\}.\quad ■$$

Exercise 1.6   Partition $\mathbb{Z}_{10}$ into cosets of the subgroup H, when

(a) $H = \{0, 2, 4, 6, 8\}$;

(b) $H = \{0, 5\}$.

Cosets in infinite additive groups

So far we have partitioned only finite groups into cosets. However, the definition of coset is applicable to *any* group. What is more, the result and proof of Theorem 1.1, which states that cosets of a subgroup form a partition of a group, makes no reference to the group being finite and so holds generally. One problem with which we shall be confronted later is that a partition of an infinite group into cosets may involve infinitely many cosets. In that case, Strategy 1.1 can be used to discover more and more cosets, but it will never lead to our finding *all* of them. In the next example, we consider a subgroup of an infinite group with a *finite* number of cosets. In such cases, Strategy 1.1 can be used to find all the cosets.

For example, we shall consider the set of cosets of $\mathbb{Z}$ in the group $\mathbb{R}$ in Subsection 6.2.

Example 1.6   Find the partition of $\mathbb{Z}$ into cosets of $3\mathbb{Z}$.

Solution   We use Strategy 1.1 for partitioning a group into cosets.

We showed that $3\mathbb{Z}$ is a subgroup of $\mathbb{Z}$ in Unit GTA2, Exercise 1.2(a).

1. We take the subgroup $3\mathbb{Z}$ as the first coset:

$$3\mathbb{Z} = \{\ldots, -6, -3, 0, 3, 6, \ldots\}.$$

2. To find another coset, we choose any element of $\mathbb{Z}$ which is not in $3\mathbb{Z}$, say 1.

We determine $1 + 3\mathbb{Z}$, the second coset:

$$\begin{aligned} 1 + 3\mathbb{Z} &= 1 + \{\ldots, -6, -3, 0, 3, 6, \ldots\} \\ &= \{\ldots, -5, -2, 1, 4, 7, \ldots\}. \end{aligned}$$

The coset $1 + 3\mathbb{Z}$ is obtained by adding 1 to each element of the subgroup $3\mathbb{Z}$.

To find a third coset, we choose an element which does not belong to either of the cosets found so far, say 2.

We determine $2 + 3\mathbb{Z}$, the third coset:

$$\begin{aligned} 2 + 3\mathbb{Z} &= 2 + \{\ldots, -6, -3, 0, 3, 6, \ldots\} \\ &= \{\ldots, -4, -1, 2, 5, 8, \ldots\}. \end{aligned}$$

The coset $2 + 3\mathbb{Z}$ is obtained by adding 2 to each element of the subgroup $3\mathbb{Z}$.

3. Every integer has been assigned to a coset, so the partition is complete. There are just three cosets of the subgroup $3\mathbb{Z}$ in $\mathbb{Z}$—namely:

$$\begin{aligned} 3\mathbb{Z} &= \{\ldots, -6, -3, 0, 3, 6, \ldots\}, \\ 1 + 3\mathbb{Z} &= \{\ldots, -5, -2, 1, 4, 7, \ldots\}, \\ 2 + 3\mathbb{Z} &= \{\ldots, -4, -1, 2, 5, 8, \ldots\}. \quad\blacksquare \end{aligned}$$

Exercise 1.7   Find:

(a) the partition of $\mathbb{Z}$ into cosets of $4\mathbb{Z}$;

(b) the partition of $2\mathbb{Z}$ into cosets of $6\mathbb{Z}$.

1.4 The video programme: 'Cosets'

The programme is a general programme about cosets, Lagrange's Theorem and normal subgroups. It uses the idea of cosets to examine the structure of the group of order 24 defined in the following table. The programme can be watched now or at any time during your study of the rest of the unit.

∘	e	a	b	c	d	f	g	h	i	j	k	l	z	y	x	w	v	u	t	s	r	q	p	o
e	e	a	b	c	d	f	g	h	i	j	k	l	z	y	x	w	v	u	t	s	r	q	p	o
a	a	e	c	b	h	g	f	d	k	l	i	j	y	z	w	x	s	t	u	v	p	o	r	q
b	b	c	e	a	f	d	h	g	l	k	j	i	w	x	y	z	t	s	v	u	q	r	o	p
c	c	b	a	e	g	h	d	f	j	i	l	k	x	w	z	y	u	v	s	t	o	p	q	r
d	d	f	g	h	i	j	k	l	e	a	b	c	p	o	r	q	w	x	y	z	u	v	s	t
f	f	d	h	g	l	k	j	i	b	c	e	a	o	p	q	r	z	y	x	w	s	t	u	v
g	g	h	d	f	j	i	l	k	c	b	a	e	q	r	o	p	y	z	w	x	v	u	t	s
h	h	g	f	d	k	l	i	j	a	e	c	b	r	q	p	o	x	w	z	y	t	s	v	u
i	i	j	k	l	e	a	b	c	d	f	g	h	s	t	u	v	q	r	o	p	x	w	z	y
j	j	i	l	k	c	b	a	e	g	h	d	f	t	s	v	u	p	o	r	q	z	y	x	w
k	k	l	i	j	a	e	c	b	h	g	f	d	v	u	t	s	o	p	q	r	w	x	y	z
l	l	k	j	i	b	c	e	a	f	d	h	g	u	v	s	t	r	q	p	o	y	z	w	x
z	z	y	x	w	u	v	s	t	q	r	o	p	e	a	b	c	f	d	h	g	j	i	l	k
y	y	z	w	x	t	s	v	u	o	p	q	r	a	e	c	b	g	h	d	f	l	k	j	i
x	x	w	z	y	v	u	t	s	p	o	r	q	c	b	a	e	h	g	f	d	i	j	k	l
w	w	x	y	z	s	t	u	v	r	q	p	o	b	c	e	a	d	f	g	h	k	l	i	j
v	v	u	t	s	p	o	r	q	x	w	z	y	k	l	i	j	e	a	b	c	g	h	d	f
u	u	v	s	t	q	r	o	p	z	y	x	w	l	k	j	i	c	b	a	e	d	f	g	h
t	t	s	v	u	o	p	q	r	y	z	w	x	j	i	l	k	b	c	e	a	h	g	f	d
s	s	t	u	v	r	q	p	o	w	x	y	z	i	j	k	l	a	e	c	b	f	d	h	g
r	r	q	p	o	w	x	y	z	s	t	u	v	h	g	f	d	l	k	j	i	e	a	b	c
q	q	r	o	p	z	y	x	w	u	v	s	t	g	h	d	f	i	j	k	l	b	c	e	a
p	p	o	r	q	x	w	z	y	v	u	t	s	d	f	g	h	j	i	l	k	a	e	c	b
o	o	p	q	r	y	z	w	x	t	s	v	u	f	d	h	g	k	l	i	j	c	b	a	e

This is a suggested point at which to watch the video 'Cosets'.

Video

Further exercises

Exercise 1.8   Let $(Q, \circ)$ be the group with the following group table.

∘	e	i	j	k	w	x	y	z
e	e	i	j	k	w	x	y	z
i	i	w	k	y	x	e	z	j
j	j	z	w	i	y	k	e	x
k	k	j	x	w	z	y	i	e
w	w	x	y	z	e	i	j	k
x	x	e	z	j	i	w	k	y
y	y	k	e	x	j	z	w	i
z	z	y	i	e	k	j	x	w

Determine the partition of Q into cosets of each of the following subgroups H.

(a) $H = \{e, w\}$ (b) $H = \{e, i, w, x\}$

Exercise 1.9 The following subgroup H of S_4 comprises the elements which fix the symbol 1:

$$H = \{e, (2\ 3), (2\ 4), (3\ 4), (2\ 3\ 4), (2\ 4\ 3)\}.$$

(a) Show that each element of the coset $(1\ 2)H$ maps the symbol 1 to the symbol 2, and hence *write down* the six elements of the coset $(1\ 2)H$. (You do not need to calculate any composites.)

(b) Using reasoning similar to that used in part (a), write down the other two cosets of H in S_4 to complete the partition of S_4 into cosets of H.

Exercise 1.10 Determine the partition of $\mathbb{Z}_{18}$ into cosets of each of the following subgroups H.

(a) $H = \{0, 9\}$ (b) $H = \{0, 6, 12\}$

Exercise 1.11 Determine the partition of $(\mathbb{Z}_{13}^*, \times_{13})$ into cosets of the subgroup $H = \{1, 12\}$.

2   Lagrange's Theorem for finite groups

After working through this section, you should be able to:

(a) understand the statement of Lagrange's Theorem;

(b) write down all possible orders for subgroups of a finite group of given order;

(c) recognise that the converse of Lagrange's Theorem is false;

(d) write down all possible orders for elements of a finite group of given order;

(e) understand the structure of all groups of prime order.

In this section we state and prove one of the most important results in group theory—Lagrange's Theorem.

2.1 Possible orders of subgroups

Each of the subgroups of S_3 that we found earlier has an order which divides 6. For example:

the subgroup $\{e\}$ has order 1;

the subgroups $\{e, (1\ 2)\}$ and $\{e, (2\ 3)\}$ have order 2;

the subgroup $\{e, (1\ 2\ 3), (1\ 3\ 2)\}$ has order 3;

the subgroup S_3 has order 6.

The group S_3 has no subgroup of order 4 or 5.

Joseph-Louis Lagrange (1736–1813) was an Italian mathematician who spent most of his working life in France. He made major contributions to number theory, mechanics, algebra and the theory of functions.
(Grateful acknowlegement is made to the Royal College of Physicians for this image of Lagrange.)

Similarly, the order of each subgroup of the group $S(\square)$ divides 8, the order of $S(\square)$. For example:

the subgroup $\{e\}$ has order 1;

the subgroups $\{e, b\}$ and $\{e, s\}$ have order 2;

the subgroups $\{e, a, b, c\}$ and $\{e, b, r, t\}$ have order 4;

the subgroup $S(\square)$ has order 8.

The group $S(\square)$ has no subgroups of order 3, 5, 6 or 7.

Likewise the order of each subgroup of the group S_4 divides 24, the order of S_4. For example:

The subgroups of S_4 are discussed in Unit GTA3, Section 4.

the subgroup $\{e\}$ has order 1;

the subgroups $\{e, (1\ 2)\}$ and $\{e, (1\ 2)(3\ 4)\}$ have order 2;

the subgroup $\{e, (1\ 2\ 3), (1\ 3\ 2)\}$ has order 3;

the subgroup $\{(e, (12)(3\ 4), (1\ 3)(2\ 4), (1\ 4)(2\ 3)\}$ has order 4;

the subgroup $\{e, (1\ 3), (1\ 4), (3\ 4), (1\ 3\ 4), (1\ 4\ 3)\}$ has order 6;

the subgroup $\{e, (1\ 3), (2\ 4), (1\ 3)(2\ 4), (1\ 2)(3\ 4), (1\ 4)(2\ 3),$
$\qquad (1\ 2\ 3\ 4), (1\ 4\ 3\ 2)\}$ has order 8;

the subgroup A_4 has order 12;

the subgroup S_4 has order 24.

In each of these examples, the order of each subgroup divides the order of the group. This is true for finite groups in general.

Theorem 2.1 Lagrange's Theorem

Let G be a finite group and let H be any subgroup of G. Then the order of H divides the order of G.

Lagrange obtained his theorem in the context of permutations of the roots of equations. At that time the concept of an abstract group had not been formulated.

For example, if G is a group of order 12, then any subgroup of G has order 1, 2, 3, 4, 6 or 12; the group G cannot have a subgroup of order 5, 7, 8, 9, 10 or 11.

Exercise 2.1 Let G be a group of order n and let H be a subgroup of G. List all the possible orders of H in each of the following cases.

(a) $n = 20$ (b) $n = 25$ (c) $n = 29$

To prove Lagrange's Theorem, we use the following properties of cosets of a subgroup H in a given finite group G.

- The cosets of H partition G.

Theorem 1.1

- Each coset gH has the same number of elements as H.

See property 1, page 7.

Proof of Lagrange's Theorem

Suppose that G has order n and that H has order m. We prove that m divides n.

Since the cosets of H partition G, each element of G lies in exactly one coset. Let the number of distinct cosets be k.

Each coset has exactly m elements, the same number as H.

Thus, as each of the k cosets has m elements, there are km elements in all.

There are k distinct cosets, each with m elements, so the total number of elements is km.

Therefore $n = km$, and m divides n, as required. ■

We illustrate this proof by taking G to be the group $S(\square)$ and H to be the subgroup $\{e, b\}$. Then there are four distinct cosets of H in G:

$$\{e, b\}, \quad \{a, c\}, \quad \{r, t\}, \quad \{s, u\}.$$

Example 1.3

Thus, in this case,

$$n = 8, \quad \text{the order of } S(\square),$$
$$k = 4, \quad \text{the number of cosets,}$$
$$m = 2, \quad \text{the order of } H,$$

and the equation $n = km$ is simply $8 = 4 \times 2$.

Exercise 2.2 Verify the equation $n = km$ when

(a) $G = S_4$ and $H = A_4$; (b) $G = \mathbb{Z}_9$ and $H = \{0, 3, 6\}$.

The number k of distinct cosets has a special name.

> **Definition** Let H be a subgroup of a group G. Then the number k of distinct cosets of H in G is called the **index** of H in G.

A subgroup H of an infinite group may have finite index; otherwise, H is said to have *infinite index*.

Remark If G is a *finite* group, the index of H in G is the order of G divided by the order of H:

$$k = |G|/|H|.$$

For example, if $H = \{e, b\}$ in $S(\square)$, then

$$\text{index of } H \text{ in } G = 8/2 = 4.$$

Exercise 2.3 Write down the index of a subgroup H in a group G, when

(a) G has order 40 and H has order 8;

(b) G has order 45 and H has order 3.

Lagrange's Theorem allows us to write down all the *possible* orders for subgroups of a finite group G—namely, all the positive divisors of the order of G. Thus, if the number m does *not* divide the order of G, then G does not have a subgroup of order m.

> **Warning** The converse of Lagrange's Theorem is *false*.
>
> Lagrange's Theorem does *not* assert that if m is a positive divisor of the order of a group G, then G has a subgroup of order m.

For example, the alternating group A_4 comprises the twelve even permutations in S_4:

$$A_4 = \{e, (1\ 2\ 3), (1\ 2\ 4), (1\ 3\ 4), (2\ 3\ 4), (1\ 3\ 2), (1\ 4\ 2),$$
$$(1\ 4\ 3), (2\ 4\ 3), (1\ 2)(3\ 4), (1\ 3)(2\ 4), (1\ 4)(2\ 3)\}.$$

The group A_4 has order 12, and the positive divisors of 12 are 1, 2, 3, 4, 6 and 12; so any subgroup of A_4 must have order 1, 2, 3, 4, 6 or 12. In fact, A_4 has subgroups of order 1, 2, 3, 4 and 12, but A_4 has no subgroup of order 6.

You are asked to show that A_4 has no subgroup of order 6 in Exercise 3.6.

Exercise 2.4   Write down a subgroup of A_4 of each of the orders 1, 2, 3, 4 and 12.

2.2 Consequences of Lagrange's Theorem

There may not seem to be anything particularly remarkable about Lagrange's Theorem. If you spent a little time searching for subgroups of given groups, you would quickly suspect the result that Lagrange discovered, even if you were unable to prove it. Nevertheless, it is a cornerstone in the theory of finite groups. We look now at some useful consequences of the result.

Cyclic subgroups

Lagrange's Theorem gives us some information about the orders of the elements of a group G because the order of an element g is the same as the order of the cyclic subgroup $\langle g \rangle$.

For example, in $S(\square)$, we have the following.

Element	Cyclic subgroup	Order
e	$\langle e \rangle = \{e\}$	1
a	$\langle a \rangle = \{e, a, a^2, a^3\}$	4
b	$\langle b \rangle = \{e, b\}$	2
c	$\langle c \rangle = \{e, c, c^2, c^3\}$	4
r	$\langle r \rangle = \{e, r\}$	2
s	$\langle s \rangle = \{e, s\}$	2
t	$\langle t \rangle = \{e, t\}$	2
u	$\langle u \rangle = \{e, u\}$	2

Note that
$$\{e, a, a^2, a^3\} = \{e, a, b, c\}$$
and
$$\{e, c, c^2, c^3\} = \{e, c, b, a\}.$$

In each case, the order of the element divides 8, the order of $S(\square)$.

In general, for any finite group G, if g is any element of order m in G, then g generates a cyclic subgroup of G:

$$\langle g \rangle = \{e, g, g^2, g^3, \ldots, g^{m-1}\}.$$

The order of the subgroup $\langle g \rangle$ is equal to m, the order of the element g. But the order of any subgroup divides the order of G, by Lagrange's Theorem. Thus m, the order of the element g, divides the order of G.

We have therefore proved the following result.

Corollary 1 to Lagrange's Theorem

Let g be an element of a finite group G. Then the order of g divides the order of G.

Exercise 2.5   Verify that the order of the element g divides the order of the group G, in each of the following cases:

(a) $G = S_4$ and $g = (1\ 2\ 3\ 4)$;   (b) $G = S_4$ and $g = (1\ 3\ 4)$;

(c) $G = \mathbb{Z}_9$ and $g = 5$;   (d) $G = \mathbb{Z}_9$ and $g = 6$.

Groups of prime order

We look next at groups of prime order. Let G be a group whose order is a prime number p. Since the only positive divisors of p are 1 and p, we deduce from Lagrange's Theorem that any subgroup of G must have order 1 or order p. But the only subgroup of order 1 is $\{e\}$ and the only subset of G with p elements is G itself. So a group of prime order cannot have any subgroups other than the identity subgroup and the whole group.

Moreover, by Corollary 1 to Lagrange's theorem, the order of each element of G is either 1 or p. But in any group, the only element of order 1 is the identity e and so all the other $p - 1$ elements must have order p. Thus G is a group of order p which has elements of order p and so G is a cyclic group.

We summarise these facts in the following result.

Corollary 2 to Lagrange's Theorem

If G is a group of prime order, then G is a cyclic group. The only subgroups of G are $\{e\}$ and G, and each element of G except e generates G.

We can now identify the structure of *any* group of prime order p. We saw earlier that any two cyclic groups of the same order are isomorphic and that the additive group $\mathbb{Z}_p$ is a cyclic group of order p. It follows that any cyclic group of order p is isomorphic to $\mathbb{Z}_p$. Combining this fact with Corollary 2 to Lagrange's theorem, we obtain the following result.

Unit GTA2, Theorem 3.1.

Unit GTA2, Theorem 4.1.

Corollary 3 to Lagrange's Theorem

If G is a group of prime order p, then G is isomorphic to the cyclic group $\mathbb{Z}_p$.

Exercise 2.6   Consider the group $G = (\{v, w, x, y, z\}, \circ)$ of order 5, defined by the following Cayley table.

$\circ$	v	w	x	y	z
v	w	z	y	v	x
w	z	x	v	w	y
x	y	v	z	x	w
y	v	w	x	y	z
z	x	y	w	z	v

(a) Why is G a cyclic group?

(b) Identify the identity element and verify that all the other elements have order 5.

(c) Find an isomorphism ϕ which maps G onto $\mathbb{Z}_5$.

Hint: Write down a generator of G and a generator of $\mathbb{Z}_5$, and then find an isomorphism by matching powers of generators (as in Unit GTA2, Strategy 3.3).

Exercise 2.7

(a) Let G be a group of order 14. Show that all the proper subgroups of G are cyclic.

(b) Let G be a group of order 15. Show that all the proper subgroups of G are cyclic.

(c) Let G be a group of order pq, where p and q are both prime. Show that all the proper subgroups of G are cyclic.

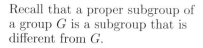

Recall that a proper subgroup of a group G is a subgroup that is different from G.

The results in Corollaries 2 and 3 to Lagrange's Theorem are very helpful when we tackle the problem of finding the number of non-isomorphic groups of a given order. For example, Corollary 3 tells us that there is just one isomorphism class for each of the prime orders

$$2, \ 3, \ 5, \ 7, \ 11, \ 13, \ 17, \ 19, \ 23, \ 29, \ 31, \ \ldots.$$

The isomorphism class for groups of prime order p comprises all the cyclic groups isomorphic to $\mathbb{Z}_p$.

The number of isomorphism classes for groups of order n for non-prime values of n is known for small values of n. But the problem of finding the isomorphism classes becomes more difficult as n increases. For example, there are five isomorphism classes for groups of order 8 and fourteen isomorphism classes for groups of order 16. We describe the isomorphism classes for groups of orders 1 to 8 in the next section.

Further exercises

Exercise 2.8

(a) For each of the orders of a group G given below, write down all the possible orders of subgroups of G.

(i) 17   (ii) 36   (iii) 56   (iv) 59

(b) In which of the above cases is G necessarily a cyclic group?

(c) In which of the above cases might G be a cyclic group?

(d) Determine the order of the subgroup of the group $\mathbb{Z}_{59}$ generated by the element 7.

Exercise 2.9   The group $S(\bigcirc)$ has order 12. A diagonal is added to the hexagon, as shown. The symmetry group of the resulting figure is a subgroup of $S(\bigcirc)$; it consists of those symmetries of the hexagon which map the diagonal to itself.

(a) What is the order of the symmetry group of the given figure?

(b) It follows from Lagrange's Theorem that the group $S(\bigcirc)$ may have proper subgroups of each of the orders 1, 2, 3, 4 and 6. Modify the regular hexagon by adding features so that the resulting figures have as their symmetry group a subgroup of $S(\bigcirc)$ with order 1, 2, 3 and 6.

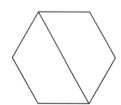

3   Groups of small order

After working through this section, you should be able to:

(a) state the number of isomorphism classes for groups of each of the orders $1, 2, \ldots, 8$;

(b) determine the isomorphism class to which a given group of order 8 or less belongs;

(c) describe the isomorphism classes for groups of orders $1, 2, \ldots, 7$.

In this section we discuss the isomorphism classes for groups of orders $1, 2, \ldots, 8$. We show that:

- there is only one isomorphism class for groups of each of the orders 1, 2, 3, 5 and 7;
- there are two isomorphism classes for groups of order 4, and two isomorphism classes for groups of order 6;
- there are five isomorphism classes for groups of order 8.

In this work we shall use the following facts.

- Isomorphic groups have the same order.   See Unit GTA2, Section 3.1.
- Cyclic groups of the same order are isomorphic.   See Unit GTA2, Theorem 3.1.
- Lagrange's Theorem: The order of a subgroup of a finite group divides the order of the group.   Theorem 2.1
- If G is a group of order n and G contains an element of order n, then G is a cyclic group.   See Unit GTA2, Subsection 2.1, Frame 18, Theorem 2.3.
- In a finite group, the order of each element divides the order of the group.   See Corollary 1 to Lagrange's Theorem.
- If G is a group of prime order p, then G is isomorphic to $\mathbb{Z}_p$.   See Corollary 3 to Lagrange's Theorem.

3.1   Groups of orders 1, 2, 3, 5 and 7

Groups of order 1

Since every group contains an identity element, a group of order 1 consists of this element alone. So if the identity is e, the Cayley table is simply the following.

$\circ$	c
e	e

We therefore make the following observation.

> **Groups of order 1**   There is only one isomorphism class for groups of order 1.

Examples of particular members of this class are

$$(\{0\}, +), \quad (\{1\}, \times) \quad \text{and} \quad S(F),$$

where F is any figure which has the identity as its only symmetry.

Groups of orders 2, 3, 5 and 7

We can deal with these groups very easily by using Corollary 3 to Lagrange's Theorem, which states that a group of prime order p is isomorphic to the cyclic group $\mathbb{Z}_p$.

Groups of orders p = 2, 3, 5 and 7 For each prime p, there is only one isomorphism class of groups of order p and all the groups in this class are cyclic.

Some familiar examples of these groups are the groups which arise in additive modular arithmetics,

$$\mathbb{Z}_2, \quad \mathbb{Z}_3, \quad \mathbb{Z}_5 \quad \text{and} \quad \mathbb{Z}_7.$$

3.2 Three useful results

Before considering groups of orders 4, 6 and 8, we prove three theorems.

Theorem 3.1 Let G be a group in which each element except the identity has order 2. Then G is Abelian.

Proof Let x and y be any two elements of G.

Since each element of G except e has order 2, we know that

$$x^2 = y^2 = (xy)^2 = e.$$

We can write the last equality in the form

$$e = (xy)(xy).$$

Multiplying both sides on the left by x and on the right by y, we get

$$
\begin{aligned}
xey &= x(xy)(xy)y \\
&= x^2(yx)y^2 \quad \text{(associativity)} \\
&= e(yx)e \quad \text{(since } x^2 = y^2 = e\text{)}.
\end{aligned}
$$

Thus, for all $x, y \in G$,

$$xy = yx \quad \text{(identity)},$$

so G is Abelian. ■

Alternatively,

$$g = g^{-1}, \text{ for each } g \in G,$$

so

$$
\begin{aligned}
(xy) &= (xy)^{-1} \\
&= y^{-1}x^{-1} \\
&= yx.
\end{aligned}
$$

Theorem 3.2 Let G be a group (with order greater than 2) in which each element except the identity has order 2. Then the order of G is a multiple of 4.

For example, $S(\square)$ satisfies these conditions, and has order 4.

Proof Let x and y be distinct non-identity elements of G; then $x \neq y$, $x \neq e$ and $y \neq e$.

The elements x and y have order 2, so $x^2 = y^2 = e$; that is, x and y are self-inverse.

Let $z = xy$; then $z \in G$, by the closure axiom G1. Also, the elements e, x, y and z are distinct:

$z \neq e$ because $z = e$ would imply $xy = e$, and thus $y = x^{-1} = x$;

$z \neq x$ because $z = x$ would imply $xy = x$, and thus $y = e$;

$z \neq y$ because $z = y$ would imply $xy = y$, and thus $x = e$.

By Theorem 3.1, the group G is Abelian, so

$yx = xy = z$.

Further,

$xz = x(xy) = x^2 y = ey = y$,   so $zx = y$ also;

$yz = y(xy) = y(yx) = y^2 x = ex = x$,   so $zy = x$ also;

$z^2 = e$ because $z \in G$.

Thus part of the group table for G is the following.

	e	x	y	z
e	e	x	y	z
x	x	e	z	y
y	y	z	e	x
z	z	y	x	e

We deduce that $\{e, x, y, z\}$ is a subgroup of G. By Lagrange's Theorem, the order of this subgroup, 4, divides the order of G, so the order of G is a multiple of 4.   ■

Theorem 3.3   Let G be a group of even order. Then G contains an element of order 2.

Proof   For each element of $g \in G$,

EITHER g is self-inverse ($g = g^{-1}$)

OR g and g^{-1} are distinct elements which are inverses of each other.

It follows that the number of elements which are not self-inverse must be even and, since $|G|$ is even, the number of elements which are self-inverse must be even. However, e is self-inverse, so there must be at least one element $g \in G$ such that

$g \circ g = e$ and $g \neq e$;

In fact, there must exist an *odd* number of elements of order 2.

that is, G contains an element of order 2.   ■

3.3 Groups of order 4

If G is a group of order 4, then, by Corollary 1 to Lagrange's Theorem, the order of each element of G divides 4. Thus each element of G has order 1, 2 or 4. The identity element is the only element of order 1.

There are two possible cases:

(a) G has an element of order 4;

(b) G has no element of order 4;

We consider these in turn.

Case (a): G has an element of order 4

If g is an element of G of order 4, then g generates G and so G is a cyclic group of order 4. It follows that G is isomorphic to C_4 and is therefore Abelian.

Case (b): G has no element of order 4

In this case, each of the three non-identity elements has order 2.

Arguing as in the proof of Theorem 3.2, we find that the group consists of four elements e, x, y and $z = xy$ with the following group table.

	e	x	y	z
e	e	x	y	z
x	x	e	z	y
y	y	z	e	x
z	z	y	x	e

So G is an Abelian group isomorphic to the Klein group K_4.

We have established the following result.

Groups of order 4   There are two isomorphism classes of groups of order 4:

> one contains the cyclic group C_4;
>
> the other contains the Klein group K_4.

All groups of order 4 are Abelian.

Remark   If you are given a group G of order 4, you can determine to which isomorphism class it belongs—just look at the orders of the elements as follows.

> If there is an element of order 4 (there would be two such elements), then $G \cong C_4$.
>
> If all elements are self-inverse, then $G \cong K_4$.

Exercise 3.1   Determine whether each of the following subgroups of S_6 is isomorphic to C_4 or to K_4.

(a) $\{e, (1\ 3), (2\ 5), (1\ 3)(2\ 5)\}$

(b) $\{e, (2\ 3\ 4\ 6), (2\ 4)(3\ 6), (2\ 6\ 4\ 3)\}$

Here we use the symbol $\cong$ for 'is isomorphic to' that we introduced in Unit GTA2, Subsection 3.1.

3.4 Groups of order 6

If G is a group of order 6, then, by Corollary 1 to Lagrange's Theorem, the order of each element of G divides 6. Thus each element of G has order 1, 2, 3 or 6. The identity element is the only element of order 1.

There are two possible cases.

(a) G has an element of order 6;

(b) G has no element of order 6.

We consider these in turn.

Case (a): G has an element of order 6

If g is an element of order 6, then g generates G and so G is a cyclic group of order 6. It follows that G is isomorphic to C_6 and is therefore Abelian.

Examples of cyclic groups of order 6 are:

$+_6$	0	1	2	3	4	5
0	0	1	2	3	4	5
1	1	2	3	4	5	0
2	2	3	4	5	0	1
3	3	4	5	0	1	2
4	4	5	0	1	2	3
5	5	0	1	2	3	4

$(\mathbb{Z}_6, +_6)$

$\times_7$	1	2	3	4	5	6
1	1	2	3	4	5	6
2	2	4	6	1	3	5
3	3	6	2	5	1	4
4	4	1	5	2	6	3
5	5	3	1	6	4	2
6	6	5	4	3	2	1

$(\mathbb{Z}_7^*, \times_7)$

The group $(\mathbb{Z}_6, +_6)$ can be generated by 1 or 5, and the group $(\mathbb{Z}_7^*, \times_7)$ can be generated by 3 or 5.

Case (b): G has no element of order 6

In this case, each non-identity element of G has order 2 or 3. Not all these elements have order 2, by Theorem 3.2, since the order of G is 6, which is not a multiple of 4.

So G must contain an element h of order 3. By Theorem 3.3, the group G must also contain an element g of order 2.

Let H be the cyclic subgroup generated by h:

$$H = \langle h \rangle = \{e, h, h^2\}.$$

Now H has order 3 and so has index 2 in G; also, $g \notin H$, by Corollary 1 to Lagrange's Theorem. Therefore we can write G as a union of cosets:

$$G = H \cup gH;$$

that is,

$$G = \{e, h, h^2\} \cup \{g, gh, gh^2\},$$

where $h^3 = e$ and $g^2 = e$.

Thus part of the group table for G looks like the following.

	e	h	h^2	g	gh	gh^2
e	e	h	h^2	g	gh	gh^2
h	h	h^2	e	?		
h^2	h^2	e	h			
g	g	gh	gh^2	e	h	h^2
gh	gh	gh^2	g			
gh^2	gh^2	g	gh			

However, to make further progress we need to identify the element hg. It cannot be g since this already appears in the same column. Similarly, it cannot be h, h^2 or e since they appear in the same row. That leaves gh or gh^2.

If $hg = gh$, then we would have

$$(hg)^2 = (hg)(hg) = (hg)(gh) = hg^2h = heh = h^2 \neq e$$

and

$$(hg)^3 = (hg)^2(hg) = (h^2)(hg) = h^3g = eg = g \neq e$$

which would mean that the order of hg is neither 2 nor 3. By Corollary 1 to Lagrange's Theorem, we would then have to conclude that hg has order 6. This would contradict our assumption that G has no elements of order 6, so $hg \neq gh$.

The only remaining possibility is that $hg = gh^2$.

There is now only one way to complete the rest of the table. If we use the observation that

$$(gh)(gh) = g(hg)h = g(gh^2)h = g^2h^3 = e,$$

and the fact that in a group table each element occurs exactly once in each row and in each column, we obtain the following table.

	e	h	h^2	g	gh	gh^2
e	e	h	h^2	g	gh	gh^2
h	h	h^2	e	gh^2	g	gh
h^2	h^2	e	h	gh	gh^2	g
g	g	gh	gh^2	e	h	h^2
gh	gh	gh^2	g	h^2	e	h
gh^2	gh^2	g	gh	h	h^2	e

Familiar examples of non-Abelian groups of order 6 are the groups $S(\triangle)$ and S_3 which have the following group tables.

$\circ$	e	a	b	r	s	t
e	e	a	b	r	s	t
a	a	b	e	t	r	s
b	b	e	a	s	t	r
r	r	s	t	e	a	b
s	s	t	r	b	e	a
t	t	r	s	a	b	e

$\circ$	e	(123)	(132)	(23)	(13)	(12)
e	e	(123)	(132)	(23)	(13)	(12)
(123)	(123)	(132)	e	(12)	(23)	(13)
(132)	(132)	e	(123)	(13)	(12)	(23)
(23)	(23)	(13)	(12)	e	(123)	(132)
(13)	(13)	(12)	(23)	(132)	e	(123)
(12)	(12)	(23)	(13)	(123)	(132)	e

Notice that, as expected, each of these two group tables has the same pattern as the group G in case (b) above. We can summarise our results as follows.

Groups of order 6   There are two isomorphism classes of groups of order 6:

> one contains the cyclic group C_6;
>
> the other contains the non-Abelian group $S(\triangle)$.

Remark   If you are given a group G of order 6, you can determine to which isomorphism class it belongs as follows.

If the group table is symmetrical about the leading diagonal, then $G \cong C_6$. G is Abelian.

If the group table is not symmetrical about the leading diagonal, then $G \cong S(\triangle)$. G is non-Abelian.

Exercise 3.2 In the symmetric group S_6, find:

(a) a subgroup isomorphic to C_6;

(b) a subgroup isomorphic to $S(\triangle)$.

3.5 Groups of order 8

If G is a group of order 8, then, by Corollary 1 to Lagrange's Theorem, the order of any element of G divides 8. Thus each element of G has order 1, 2, 4 or 8. The identity element is the only element of order 1.

There are three possible cases.

(a) G has an element of order 8.

(b) Each element of G except e has order 2.

(c) G has no element of order 8, but has an element of order 4.

We discuss these three cases briefly, omitting the proofs.

Case (a): G has an element of order 8

If g is an element of G of order 8, then g generates G and so G is a cyclic group of order 8. It follows that G is isomorphic to C_8 and is therefore Abelian.

G comprises the identity, four elements of order 8, two elements of order 4 and one element of order 2.

Case (b): Each element of G except e has order 2

Here G is a group of order 8 in which each element except the identity has order 2, so, by Theorem 3.1, G is Abelian. It can be shown that all such groups are isomorphic.

An example of a group with this structure is the symmetry group of a cuboid with no square faces. (Unit GTA1, Exercise 5.4.)

G comprises the identity and seven elements of order 2.

Case (c): G has no element of order 8, but has an element of order 4

In this case, each non-identity element of G has order 2 or 4.

Using an approach similar to that used for groups of order 6, we can show that there are three non-isomorphic groups of this type:

a non-cyclic Abelian group comprising the identity, four elements of order 4 and three elements of order 2;

a non-Abelian group comprising the identity, two elements of order 4 and five elements of order 2;

a non-Abelian group comprising the identity, six elements of order 4 and one element of order 2.

These groups of order 8 can be distinguished by whether they are Abelian and by the number of elements of order 2 that they possess. We can summarise these results as follows.

Groups of order 8   There are five isomorphism classes of groups of order 8.

Abelian groups

Class 1   contains the cyclic groups; these have only *one* element of order 2.

Class 2   contains groups in which *all the non-identity* elements have order 2.

Class 3   contains groups with only *three* elements of order 2.

Non-Abelian groups

Class 4   contains groups with *five* elements of order 2.

Class 5   contains groups with only *one* element of order 2.

$S(\square)$ belongs to class 4.

This leads to the following strategy.

Strategy 3.1   To determine the isomorphism class of a group G of order 8.

1.   Determine whether G is Abelian.

2.   Find the number of elements of G which have order 2.

3.   Identify the class from the following table.

Is G Abelian?	Number of elements of order 2	Class
Yes	1	1
Yes	7	2
Yes	3	3
No	5	4
No	1	5

Exercise 3.3   Use Strategy 3.1 to determine the isomorphism class to which each of the following groups of order 8 belongs.

(a)

∘	e	a	b	c	d	f	g	h
e	e	a	b	c	d	f	g	h
a	a	e	c	b	f	d	h	g
b	b	c	e	a	g	h	d	f
c	c	b	a	e	h	g	f	d
d	d	f	g	h	e	a	b	c
f	f	d	h	g	a	e	c	b
g	g	h	d	f	b	c	e	a
h	h	g	f	d	c	b	a	e

(b)

∘	e	a	b	c	d	f	g	h
e	e	a	b	c	d	f	g	h
a	a	b	c	e	f	g	h	d
b	b	c	e	a	g	h	d	f
c	c	e	a	b	h	d	f	g
d	d	f	g	h	e	a	b	c
f	f	g	h	d	a	b	c	e
g	g	h	d	f	b	c	e	a
h	h	d	f	g	c	e	a	b

(c)

∘	e	a	b	c	d	f	g	h
e	e	a	b	c	d	f	g	h
a	a	b	c	d	f	g	h	e
b	b	c	d	f	g	h	e	a
c	c	d	f	g	h	e	a	b
d	d	f	g	h	e	a	b	c
f	f	g	h	e	a	b	c	d
g	g	h	e	a	b	c	d	f
h	h	e	a	b	c	d	f	g

(d)

∘	e	a	b	c	d	f	g	h
e	e	a	b	c	d	f	g	h
a	a	b	c	e	f	g	h	d
b	b	c	e	a	g	h	d	f
c	c	e	a	b	h	d	f	g
d	d	h	g	f	b	a	e	c
f	f	d	h	g	c	b	a	e
g	g	f	d	h	e	c	b	a
h	h	g	f	d	a	e	c	b

3.6 Summary of isomorphism classes

We conclude this section by summarising the results of the previous subsections. In the following table we list the isomorphism classes for groups of orders 1 to 7 and assign some of the groups that we have met to the appropriate classes.

Order	Representative	Structure	Some examples
1	C_1	cyclic	$(\{0\}, +)$, $(\{1\}, \times)$
2	C_2	cyclic	$(\mathbb{Z}_2, +_2)$, $S^+(\square)$, $(\mathbb{Z}_3^*, \times_3)$
3	C_3	cyclic	$(\mathbb{Z}_3, +_3)$, $S^+(\triangle)$, $(\{0, 4, 8\}, +_{12})$, $(\{1, 4, 7\}, \times_9)$
4	C_4	cyclic	$(\mathbb{Z}_4, +_4)$, $(\mathbb{Z}_5^*, \times_5)$, $S^+(\square)$, $S(\text{WIND})$, $(\{0, 3, 6, 9\}, +_{12})$, $(\{1, 9, 13, 17\}, \times_{20})$
4	K_4	non-cyclic, Abelian	$S(\square)$, $(\{1, 3, 5, 7\}, \times_8)$ $(\{1, 5, 7, 11\}, \times_{12})$, $(\{1, 4, 11, 14\}, \times_{15})$, $(\{1, 7, 9, 15\}, \times_{16})$, $(\{1, 9, 11, 19\}, \times_{20})$
5	C_5	cyclic	$(\mathbb{Z}_5, +_5)$, $S^+(\pentagon)$
6	C_6	cyclic	$(\mathbb{Z}_6, +_6)$, $S^+(\hexagon)$, $(\mathbb{Z}_7^*, \times_7)$, $(\{1, 2, 4, 5, 7, 8\}, \times_9)$, $(\{0, 2, 4, 6, 8, 10\}, +_{12})$, $(\{1, 3, 5, 9, 11, 13\}, \times_{14})$
6	$S(\triangle)$	non-cyclic, non-Abelian	$S(\triangle)$, S_3, $\{e, (2\,3), (2\,4), (3\,4), (2\,3\,4), (2\,4\,3)\}$
7	C_7	cyclic	$(\mathbb{Z}_7, +_7)$, $S^+(\text{HEPT})$

Further exercises

Exercise 3.4   Let $(P, \circ)$ and $(Q, \circ)$ be the groups with the following Cayley tables.

$\circ$	e	a	b	c	p	q	r	s
e	e	a	b	c	p	q	r	s
a	a	b	c	e	q	r	s	p
b	b	c	e	a	r	s	p	q
c	c	e	a	b	s	p	q	r
p	p	q	r	s	e	a	b	c
q	q	r	s	p	a	b	c	e
r	r	s	p	q	b	c	e	a
s	s	p	q	r	c	e	a	b

$(P, \circ)$

$\circ$	e	i	j	k	w	x	y	z
e	e	i	j	k	w	x	y	z
i	i	w	k	y	x	e	z	j
j	j	z	w	i	y	k	e	x
k	k	j	x	w	z	y	i	e
w	w	x	y	z	e	i	j	k
x	x	e	z	j	i	w	k	y
y	y	k	e	x	j	z	w	i
z	z	y	i	e	k	j	x	w

$(Q, \circ)$

Determine the isomorphism class of each of the following groups:

(a) $(P, \circ)$;   (b) the subgroup $\{e, a, b, c\}$ of $(P, \circ)$;

(c) $(Q, \circ)$;   (d) the subgroup $\{e, i, w, x\}$ of $(Q, \circ)$.

Exercise 3.5   Write down a subgroup of the group $(P, \circ)$ in Exercise 3.4 that is isomorphic to K_4.

K_4 is the Klein group.

Exercise 3.6   The alternating group A_4 comprises the 12 elements:

$$\{e, (1\ 2\ 3), (1\ 2\ 4), (1\ 3\ 4), (2\ 3\ 4), (1\ 3\ 2), (1\ 4\ 2),$$
$$(1\ 4\ 3), (2\ 4\ 3), (1\ 2)(3\ 4), (1\ 3)(2\ 4), (1\ 4)(2\ 3)\}.$$

Show that A_4 has no subgroup of order 6.

4   Normal subgroups

After working through this section, you should be able to:

(a) explain the meaning of the term *right coset*;

(b) determine the right cosets Hg of a given subgroup H in a given group G;

(c) explain what is meant by a *normal* subgroup;

(d) determine whether a given subgroup is a normal subgroup by comparing left and right partitions.

4.1 Right cosets

Let $(G, \circ)$ be a group and let H be a subgroup of G. In Section 1 we defined the coset gH, where g is an element of G; it is obtained by composing each element of H with g on the *left*:

$$gH = \{g \circ h : h \in H\}.$$

We saw that the cosets of a subgroup *partition* the group G—if we list the distinct cosets, then every element of G belongs to exactly one coset.

There was no special reason why we composed each element of H on the *left* by g; we could equally well have formed cosets by composing each element with g on the *right*.

> **Definition**   Let H be a subgroup of a group $(G, \circ)$ and let g be an element of G. The **right coset** Hg is the set of elements of G of the form $h \circ g$, where $h \in H$. That is,
>
> $$Hg = \{h \circ g : h \in H\}.$$
>
> It is the set obtained by composing each element of H with g on the right.

If H is finite, say
$$H = \{h_1, h_2, \ldots, h_m\},$$
then
$$Hg = \{h_1 \circ g, h_2 \circ g, \ldots, h_m \circ g\}.$$

Example 4.1   Determine all the distinct right cosets of the subgroup $H = \{e, r\}$ in $S(\square)$.

Solution   The group $S(\square)$ has eight elements, so there are eight potential right cosets—namely,

$$He, \quad Ha, \quad Hb, \quad Hc, \quad Hr, \quad Hs, \quad Ht, \quad Hu,$$

but if we expect these to partition $S(\square)$, then we should expect only four distinct cosets.

Now

$$He = \{e, r\}e = \{e \circ e, r \circ e\} = \{e, r\},$$
$$Ha = \{e, r\}a = \{e \circ a, r \circ a\} = \{a, u\},$$
$$Hb = \{e, r\}b = \{e \circ b, r \circ b\} = \{b, t\},$$
$$Hc = \{e, r\}c = \{e \circ c, r \circ c\} = \{c, s\},$$
$$Hr = \{e, r\}r = \{e \circ r, r \circ r\} = \{r, e\},$$
$$Hs = \{e, r\}s = \{e \circ s, r \circ s\} = \{s, c\},$$
$$Ht = \{e, r\}t = \{e \circ t, r \circ t\} = \{t, b\},$$
$$Hu = \{e, r\}u = \{e \circ u, r \circ u\} = \{u, a\}.$$

Thus there are four distinct right cosets of H in $S(\square)$:

$$He = Hr = \{e, r\},$$
$$Ha = Hu = \{a, u\},$$
$$Hb = Ht = \{b, t\},$$
$$Hc = Hs = \{c, s\}. \quad \blacksquare$$

Remarks

1. Strictly speaking, the sets that we have been calling *cosets* up to this point should have been called *left cosets*, to distinguish them from *right cosets*.

    From now onwards we shall distinguish between left and right cosets when it is necessary.

2. Notice that the right coset Ha of Example 4.1 is not the same set as the corresponding left coset aH:
    $$Ha = \{e, r\}a = \{e \circ a, r \circ a\} = \{a, u\},$$
    whereas
    $$aH = a\{e, r\} = \{a \circ e, a \circ r\} = \{a, s\}.$$
    So left and right cosets are, in general, different sets. However, in some cases the left and right cosets are the same.

3. If G is an *additive* group, we denote the right coset of a subgroup H
    in G by $H + g$. Thus

$$H + g = \{h + g : h \in H\}.$$

Again, we write the $+$ to remind ourselves that the group operation is addition.

Example 4.2   Determine all the distinct right cosets of the subgroup $H = \{e, (1\ 2\ 3), (1\ 3\ 2)\}$ in S_3.

Solution   The group S_3 has six elements, so there are six potential right cosets:

$$He, \quad H(1\ 2), \quad H(1\ 3), \quad H(2\ 3), \quad H(1\ 2\ 3), \quad H(1\ 3\ 2).$$

We expect two distinct cosets.

Now

$$\begin{aligned}
He &= \{e \circ e, (1\ 2\ 3) \circ e, (1\ 3\ 2) \circ e\} \\
&= \{e, (1\ 2\ 3), (1\ 3\ 2)\} = H, \\
H(1\ 2) &= \{e \circ (1\ 2), (1\ 2\ 3) \circ (1\ 2), (1\ 3\ 2) \circ (1\ 2)\} \\
&= \{(1\ 2), (1\ 3), (2\ 3)\}, \\
H(1\ 3) &= \{e \circ (1\ 3), (1\ 2\ 3) \circ (1\ 3), (1\ 3\ 2) \circ (1\ 3)\} \\
&= \{(1\ 3), (2\ 3), (1\ 2)\}, \\
H(2\ 3) &= \{e \circ (2\ 3), (1\ 2\ 3) \circ (2\ 3), (1\ 3\ 2) \circ (2\ 3)\} \\
&= \{(2\ 3), (1\ 2), (1\ 3)\}, \\
H(1\ 2\ 3) &= \{e \circ (1\ 2\ 3), (1\ 2\ 3) \circ (1\ 2\ 3), (1\ 3\ 2) \circ (1\ 2\ 3)\} \\
&= \{(1\ 2\ 3), (1\ 3\ 2), e\} = H, \\
H(1\ 3\ 2) &= \{e \circ (1\ 3\ 2), (1\ 2\ 3) \circ (1\ 3\ 2), (1\ 3\ 2) \circ (1\ 3\ 2)\} \\
&= \{(1\ 3\ 2), e, (1\ 2\ 3)\} = H.
\end{aligned}$$

Thus there are two distinct right cosets of H in S_3:

$$\begin{aligned}
He = H(1\ 2\ 3) = H(1\ 3\ 2) &= H, \\
H(1\ 2) = H(1\ 3) = H(2\ 3) &= \{(1\ 2), (1\ 3), (2\ 3)\}. \quad \blacksquare
\end{aligned}$$

Our decision to work with left cosets in Section 1 was quite arbitrary. If we had chosen to work with right cosets instead, then our results would have been essentially the same. In particular, we would have obtained the following analogues of the properties of the cosets listed in Subsection 1.1, each of which is illustrated in the above examples.

Properties of right cosets

1. For each element g and each subgroup H of a *finite* group, the right coset Hg has the same number of elements as H.

2. For each element g and each subgroup H, the element g lies in the right coset Hg.

3. One of the right cosets Hg is H itself.

4. Any two right cosets Hg_1 and Hg_2 are either the same set or are disjoint.

Remark   The proofs of these statements are analogues of those in Section 1.

Similarly, an analogous strategy to that for partitioning a finite group G into left cosets gH of a given subgroup H can be used to partition G into right cosets Hg.

Strategy 4.1   To partition a finite group G into right cosets of a given subgroup H.

1. Take H as the first coset.

2. Choose any element g not yet assigned to a right coset and determine the right coset Hg to which g belongs.

3. Repeat step 2 until every element of G has been assigned to a right coset.

Property 1 of right cosets states that all cosets of H have the same number of elements, so if $|G| = n$ and $|H| = m$, this process will end after we have found n/m right cosets.

Exercise 4.1   For each of the following subgroups H of $S(\square)$, determine all the right cosets.

(a) $H = \{e, s\}$ (b) $H = \{e, b, r, t\}$

Exercise 4.2   Determine all the right cosets of the subgroup $H = \{e, (1\ 2)\}$ in S_3.

Exercise 4.3   Let H be a subgroup of a finite group G and let $g \in G$. Show that

$$\text{if } g \in H, \quad \text{then} \quad Hg = H,$$

and, conversely,

$$\text{if } Hg = H, \quad \text{then} \quad g \in H.$$

This result also holds for infinite groups, but we do not prove it here.

We now have two ways to partition a group into cosets of a subgroup: a partition into *left cosets* (which we call the *left partition*) and a partition into *right cosets* (the *right partition*). We have already seen that these two partitions may not be the same. In Example 1.1 we found that the left partition of $S(\square)$ into cosets of the subgroup $\{e, r\}$ is

$$\{e, r\}, \quad \{a, s\}, \quad \{b, t\}, \quad \{c, u\},$$

whereas the right partition, which we found in Example 4.1, is

$$\{e, r\}, \quad \{a, u\}, \quad \{b, t\}, \quad \{c, s\}.$$

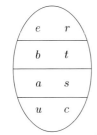

partition of $S(\square)$
into left cosets of $\{e, r\}$

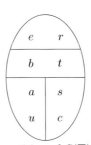

partition of $S(\square)$
into right cosets of $\{e, r\}$

Similarly, the right partition of S_3 into cosets of $\{e, (1\ 2)\}$, which we found in Exercise 4.2, is

$$\{e, (1\ 2)\}, \quad \{(1\ 3), (1\ 3\ 2)\}, \quad \{(2\ 3), (1\ 2\ 3)\},$$

and this is not the same as the left partition, which we found in Exercise 1.2, namely

$$\{e, (1\ 2)\}, \quad \{(1\ 3), (1\ 2\ 3)\}, \quad \{(2\ 3), (1\ 3\ 2)\}.$$

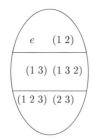 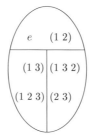

partition of S_3 into   partition of S_3 into
left cosets of $\{e, (1\ 2)\}$ right cosets of $\{e, (1\ 2)\}$

Example 4.3   Determine whether the left and right partitions of $S(\square)$ into cosets of the subgroup $H = \{e, b\}$ are the same.

Solution   We have already found the left partition in Example 1.3:

$$\{e, b\}, \quad \{a, c\}, \quad \{r, t\}, \quad \{s, u\}. \tag{4.1}$$

We now apply Strategy 4.1 for partitioning $S(\square)$ into right cosets of H.

We take the subgroup $\{e, b\}$ as the first right coset.

We choose any element not in $\{e, b\}$, say a.

We determine Ha, the second right coset:

$$\{e, b\}a = \{e \circ a, b \circ a\} = \{a, c\}.$$

We choose any element not in H or Ha, say r.

We determine Hr, the third right coset:

$$\{e, b\}r = \{e \circ r, b \circ r\} = \{r, t\}.$$

We choose any element not in H, Ha or Hr, say s.

We determine Hs, the fourth right coset:

$$\{e, b\}s = \{e \circ s, b \circ s\} = \{s, u\}.$$

Every element of $S(\square)$ has now been assigned to a coset, so the right partition is

$$\{e, b\}, \quad \{a, c\}, \quad \{r, t\}, \quad \{s, u\}. \tag{4.2}$$

The two partitions (4.1) and (4.2) are the same.   ■

Exercise 4.4   Determine whether the left and right partitions of $S(\square)$ into cosets are the same for each of the following subgroups H.

(a) $H = \{e, a, b, c\}$ (b) $H = \{e\}$ (c) $H = S(\square)$

4.2 Definition of a normal subgroup

We have seen in Examples 1.1 and 4.1 an example of a subgroup for which the left and right partitions differ, and we have seen in Examples 1.3 and 4.3 an example of a subgroup for which the left and right partitions are the same. Subgroups with the property that the two partitions are the same are special and play an important role in the theory of groups. They merit their own name.

Definition　Let G be a group and let H be a subgroup of G. Then H is a **normal subgroup** of G if the left and right partitions of G into cosets of H are the same. That is, H is **normal in** G if, for each element $g \in G$,
$$gH = Hg.$$

Often we shall use the letter N to denote a *normal* subgroup, rather than H.

For any $g \in G$, the left coset containing g is gH and the right coset containing g is Hg. If the left and right partitions of G into cosets are the same, then we must have $gH = Hg$.

For example,

in $S(\square)$,

the subgroup $\{e, b\}$ is normal,
the subgroup $\{e, r\}$ is not normal;

Example 1.3 and Example 4.3.
Example 1.1 and Example 4.1.

in S_3,

the subgroup $\{e, (1\ 2\ 3), (1\ 3\ 2)\}$ is normal,
the subgroup $\{e, (1\ 2)\}$ is not normal.

Example 1.2 and Example 4.2.
Exercise 1.2 and Exercise 4.2.

Remarks

1.　The condition $gH = Hg$ means that the sets gH and Hg contain the same elements; it does not mean that $gh = hg$ for all $h \in H$.

For example, for the subgroup $H = \{e, (1\ 2\ 3), (1\ 3\ 2)\}$ of S_3, we have
$$(1\ 2)H = \{(1\ 2), (2\ 3), (1\ 3)\},$$
$$H(1\ 2) = \{(1\ 2), (1\ 3), (2\ 3)\}.$$

These *sets* are the same, so $(1\ 2)H = H(1\ 2)$, but
$$(1\ 2) \circ (1\ 2\ 3) = (2\ 3), \quad \text{whereas} \quad (1\ 2\ 3) \circ (1\ 2) = (1\ 3).$$

2.　We use the phrases 'H is a normal subgroup of G' and 'H is normal in G' interchangeably.

Some texts use the notation
$$H \lhd G$$
to denote that H is a normal subgroup of G.

Exercise 4.5　Consider the alternating group A_4 with elements

$$e, \quad (1\ 2\ 3), \quad (1\ 3\ 2), \quad (1\ 4)(2\ 3),$$
$$(2\ 4\ 3), \quad (1\ 3\ 4), \quad (1\ 4\ 2), \quad (1\ 3)(2\ 4),$$
$$(2\ 3\ 4), \quad (1\ 4\ 3), \quad (1\ 2\ 4), \quad (1\ 2)(3\ 4).$$

(a) Let H be the subgroup $\{e, (1\ 2\ 3), (1\ 3\ 2)\}$. By calculating the cosets $(1\ 2)(3\ 4)H$ and $H(1\ 2)(3\ 4)$, show that H is not a normal subgroup of A_4.

(b) Let N be the subgroup $\{e, (1\ 2)(3\ 4), (1\ 3)(2\ 4), (1\ 4)(2\ 3)\}$. Determine the left and right partitions of N in A_4 and hence show that N is a normal subgroup of A_4.

Every group has at least two normal subgroups, as stated in the following theorem.

Theorem 4.1 Let G be a group; then

(a) the identity subgroup $\{e\}$ is a normal subgroup of G;

(b) the whole group G is a normal subgroup of G.

Exercise 4.6

(a) Prove Theorem 4.1(a) by showing that, for any $g \in G$,

$$g\{e\} = \{e\}g.$$

(b) Prove Theorem 4.1(b) by showing that, for any $g \in G$,

$$gG = Gg.$$

For some groups, the subgroups in Theorem 4.1 are the *only* normal subgroups. On the other hand, there are groups in which *every* subgroup is normal; in particular, Abelian groups have this property.

Theorem 4.2   In an Abelian group, every subgroup is normal.

Proof   Let H be any subgroup of the Abelian group G and let g be any element of G. Then the Abelian property tells us that $gh = hg$ for every pair of elements $g, h \in G$, so

$$gH = \{gh : h \in H\} = \{hg : h \in H\} = Hg.$$

Thus the left and right cosets of H in G are the same, so H is a normal subgroup of G.   ∎

Another important example of a normal subgroup arises when G is a finite group of even order, say $2m$, and H is a subgroup of order m; in other words, when the order of H is exactly *half* that of G, or, in the language of Section 2, H is a subgroup of *index 2* in G. Then, as we shall show, H must be a normal subgroup of G.

Recall from page 17 that the index of H in G is the number of distinct cosets of H in G. In the case of finite groups, the index is equal to the order of G divided by the order of H.

Exercise 4.7   Determine the two left cosets and the two right cosets of H in G for the following examples, and confirm that H is normal in each case.

(a) $G = S(\square)$, $H = \{e, b, s, u\}$.

(b) $G = \mathbb{Z}_8$, $H = \{0, 2, 4, 6\}$.

(c) $G = S_3$, $H = A_3$.

Theorem 4.3   Let H be a subgroup of index 2 in a group G. Then H is a normal subgroup of G.

Proof   The subgroup H of index 2 has just two left cosets. One left coset is the subgroup H itself and, because cosets are disjoint, the other left coset consists of all the elements of G which are not in H.

Similarly, there are just two right cosets. One right coset is the subgroup H itself and the other right coset consists of all the elements of G which are not in H. (See the following diagram.)

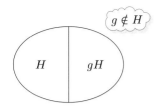
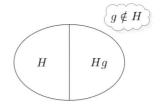

partition of G into left cosets   partition of G into right cosets

So the left and right partitions have the same two cosets and therefore H is normal in G. ■

Exercise 4.7(c) illustrates a general result. We have seen that there are just as many even permutations as there are odd permutations in S_n, so the order of A_n is $\frac{1}{2}(n!)$ and hence A_n is a subgroup of index 2. It follows from Theorem 4.3 that A_n is normal in S_n.

Unit GTA3, Theorem 2.4.

> **Corollary** For all $n \geq 2$, the alternating group A_n is a normal subgroup of the symmetric group S_n.

Further exercises

Exercise 4.8 Let $(Q, \circ)$ be the group with the following group table.

This is the same group as in Exercise 1.8.

$\circ$	e	i	j	k	w	x	y	z
e	e	i	j	k	w	x	y	z
i	i	w	k	y	x	e	z	j
j	j	z	w	i	y	k	e	x
k	k	j	x	w	z	y	i	e
w	w	x	y	z	e	i	j	k
x	x	e	z	j	i	w	k	y
y	y	k	e	x	j	z	w	i
z	z	y	i	e	k	j	x	w

Determine the partition of $(Q, \circ)$ into right cosets of each of the following subgroups H.

(a) $H = \{e, w\}$ (b) $H = \{e, i, w, x\}$

Use the partitions into left cosets that you found in Exercise 1.8 to determine whether each of these subgroups is normal in Q.

Exercise 4.9 The following subgroup H of S_4 comprises the elements which fix the symbol 1:

$$H = \{e, (2\ 3), (2\ 4), (3\ 4), (2\ 3\ 4), (2\ 4\ 3)\}.$$

(a) Show that the right coset $H(1\ 2)$ consists of the six elements in S_4 which map the symbol 2 to the symbol 1.

(b) Write down the right partition of S_4 into cosets of H.

(c) Use the result of Exercise 1.9 to decide whether H is a normal subgroup of S_4.

Exercise 4.10   In each of the following cases, the subgroup H is normal in the group G. Give a reason why this is so.

(a) $G = \mathbb{Z}_{16}$, $H = \{0, 4, 8, 12\}$.

(b) $G = S_5$, $H = A_5$.

(c) $G = (Q, \circ)$, $H = \{e, k, w, z\}$.

Here $(Q, \circ)$ is the group given in Exercise 4.8.

5   Quotient groups

After working through this section, you should be able to:

(a) combine subsets of a group using set composition;

(b) understand that *set composition* is a closed binary operation on the set of cosets of a normal subgroup of a group;

(c) explain what is meant by a *quotient group*;

(d) construct quotient groups in some simple cases.

5.1   Constructing quotient groups

In this audio section we use a binary operation · for combining subsets of a group. This operation is called *set composition*, which we define formally as follows.

Definition   Let A and B be subsets of a group $(G, \circ)$. Then the binary operation ·, called **set composition**, is defined by

$$A \cdot B = \{a \circ b : a \in A, b \in B\};$$

that is, $A \cdot B$ is the subset of G obtained by composing each element of A with each element of B on the right.

Remark   Our interest in set composition is confined to the case in which the sets A and B are *cosets* of a normal subgroup.

Example 5.1   Let H, aH, bH and cH be the cosets of the subgroup $H = \{e, t\}$ in $S(\square)$:

$$H = \{e, t\}, \quad aH = \{a, u\}, \quad bH = \{b, r\}, \quad cH = \{c, s\}.$$

Determine the following subsets of $S(\square)$.

(a) $H \cdot bH$ (b) $aH \cdot cH$

Solution

(a) $H \cdot bH = \{e, t\} \cdot \{b, r\}$

$\qquad\qquad = \{e \circ b, e \circ r, t \circ b, t \circ r\}$

$\qquad\qquad = \{b, r, r, b\} = \{r, b\}$

(b) $aH \cdot cH = \{a, u\} \cdot \{c, s\}$

$\qquad\qquad = \{a \circ c, a \circ s, u \circ c, u \circ s\}$

$\qquad\qquad = \{e, t, r, b\}$ ∎

As explained in Unit I2, Section 1.2, a set does not have repeated elements, and the elements may be written in any order.

Exercise 5.1 Let H, aH, rH and sH be the cosets of the subgroup $H = \{e, b\}$ in $S(\square)$:

$$H = \{e, b\}, \quad aH = \{a, c\}, \quad rH = \{r, t\}, \quad sH = \{s, u\}.$$

Determine the following subsets of $S(\square)$.

(a) $rH \cdot H$ (b) $sH \cdot rH$ (c) $aH \cdot aH$

Exercise 5.2 Let H, aH, bH and cH be the cosets of the subgroup $H = \{e, r\}$ in $S(\square)$:

$$H = \{e, r\}, \quad aH = \{a, s\}, \quad bH = \{b, t\}, \quad cH = \{c, u\}.$$

Determine the following subsets of $S(\square)$.

(a) $aH \cdot aH$ (b) $bH \cdot cH$ (c) $cH \cdot bH$

$\circ$	e	a	b	c	r	s	t	u
e	e	a	b	c	r	s	t	u
a	a	b	c	e	s	t	u	r
b	b	c	e	a	t	u	r	s
c	c	e	a	b	u	r	s	t
r	r	u	t	s	e	c	b	a
s	s	r	u	t	a	e	c	b
t	t	s	r	u	b	a	e	c
u	u	t	s	r	c	b	a	e

$$S(\square)$$

Next we look at an example where the group operation is additive. Here it makes sense to write $+$ rather than $\cdot$ for set composition.

We write $+$ to remind ourselves that the group operation is additive.

Example 5.2   Consider the group $\mathbb{Z}_6$ and the subgroup $H = \{0, 3\}$. The (left) cosets of H in $\mathbb{Z}_6$ are

$$H = \{0, 3\}, \quad 1 + H = \{1, 4\}, \quad 2 + H = \{2, 5\}.$$

Determine the following subsets of $\mathbb{Z}_6$.

(a) $(1 + H) + (1 + H)$ (b) $(1 + H) + (2 + H)$

Solution

(a) $(1 + H) + (1 + H) = \{1, 4\} + \{1, 4\}$
$$= \{1 +_6 1, 1 +_6 4, 4 +_6 1, 4 +_6 4\}$$
$$= \{2, 5, 5, 2\}$$
$$= \{2, 5\} = 2 + H$$

(b) $(1 + H) + (2 + H) = \{1, 4\} + \{2, 5\}$
$$= \{1 +_6 2, 1 +_6 5, 4 +_6 2, 4 +_6 5\}$$
$$= \{3, 0, 0, 3\}$$
$$= \{0, 3\} = H \quad \blacksquare$$

Exercise 5.3 For the cosets in Example 5.2, determine the following.

(a) $H + H$ (b) $(2 + H) + (1 + H)$ (c) $(2 + H) + (2 + H)$

Listen to the audio as you work through the frames.

Audio

39

1. Blocking the group table for S(□)

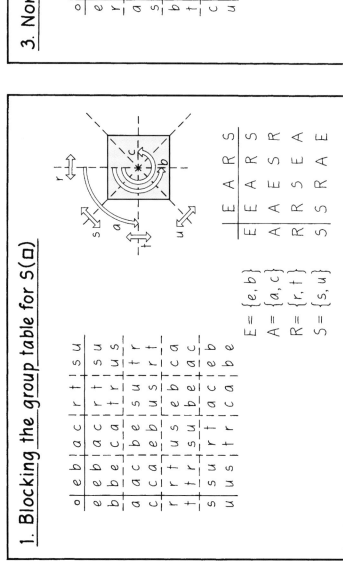

	e	b	a	c	r	t	s	u
e	e	b	a	c	r	t	s	u
b	b	e	c	a	t	r	u	s
a	a	c	e	b	s	u	r	t
c	c	a	b	e	u	s	t	r
r	r	t	u	s	e	b	c	a
t	t	r	s	u	b	e	a	c
s	s	u	r	t	c	a	e	b
u	u	s	t	r	a	c	b	e

$E = \{e, b\}$
$A = \{a, c\}$
$R = \{r, t\}$
$S = \{s, u\}$

	E	A	R	S
E	E	A	R	S
A	A	E	S	R
R	R	S	E	A
S	S	R	A	E

2. A table of cosets of subgroup H in S(□)

$H = \{e, b\} = E$
$aH = \{a, c\} = A$
$rH = \{r, t\} = R$
$sH = \{s, u\} = S$

	H	aH	rH	sH
H	H	aH	rH	sH
aH	aH	H	sH	rH
rH	rH	sH	H	aH
sH	sH	rH	aH	H

(bubble) $aH = \{a, c\} = cH$,
$rH = \{r, t\} = tH$,
...

3. Non-blocking of group table for S(□)

*(bubble: try a **different** subgroup $H = \{e, r\}$)*

(bubble: contains c, u, s, a)

	e	r	a	s	b	t	c	u
e	e	r	a	s	b	t	c	u
r	r	e	s	a	t	b	u	c
a	a	s	b	t	c	u	e	r
s	s	a	t	b	u	c	r	e
b	b	t	c	u	e	r	a	s
t	t	b	u	c	r	e	s	a
c	c	u	e	r	a	s	b	t
u	u	c	r	e	s	a	t	b

$H = \{e, r\}$
$aH = \{a, s\}$
$bH = \{b, t\}$
$cH = \{c, u\}$

	H	aH	bH	cH
H	H	*	bH	*
aH	aH	*	cH	*
bH	bH	*	H	*
cH	cH	*	aH	*

(bubble: starred entries are NOT cosets)

4. Exercise 5.4

In S(□), show that:

(a) the group table does not block for $H = \{e, s\}$;

(b) the group table does block for $H = \{e, b, r, t\}$.

5. Combining cosets of H = {e, b}

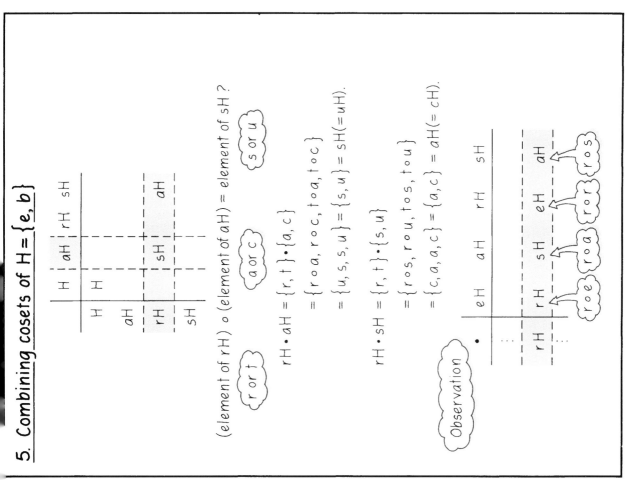

	H	aH	rH	sH
H	H			
aH				
rH			sH	aH
sH				

(element of rH) ∘ (element of aH) = element of sH?

(r or t)   *(a or c)*   *(s or u)*

$rH \cdot aH = \{r, t\} \cdot \{a, c\}$
$= \{r \circ a, r \circ c, t \circ a, t \circ c\}$
$= \{u, s, s, u\} = \{s, u\} = sH(= uH).$

$rH \cdot sH = \{r, t\} \cdot \{s, u\}$
$= \{r \circ s, r \circ u, t \circ s, t \circ u\}$
$= \{c, a, a, c\} = \{a, c\} = aH(= cH).$

• **Observation**

...	eH	aH	rH	sH
rH	sH	eH	aH	

(r∘e) *(r∘a)* *(r∘r)* *(r∘s)*

6. Combining cosets of H = {e, r}

$bH \cdot aH = \{b, t\} \cdot \{a, s\}$
$= \{b \circ a, b \circ s, t \circ a, t \circ s\}$
$= \{c, u, s, a\}$

(not a coset)

7. The general case

Let $(G, \circ)$ be any group with subgroup H.

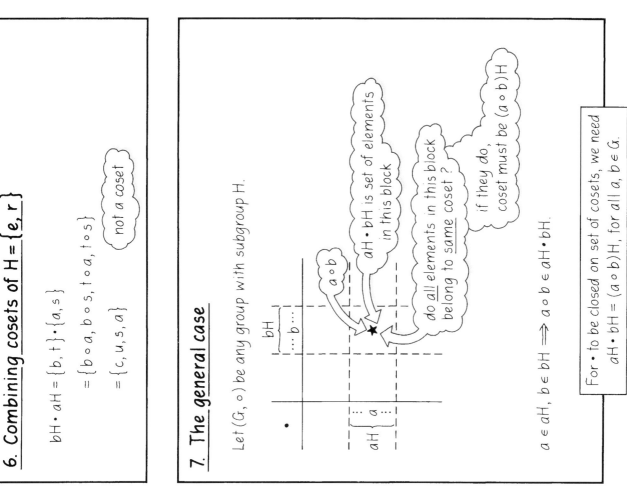

(a∘b)
(aH · bH is set of elements in this block)
(do all elements in this block belong to same coset?)
(if they do, coset must be (a∘b)H)

$a \in aH, \; b \in bH \implies a \circ b \in aH \cdot bH.$

For • to be closed on set of cosets, we need
$aH \cdot bH = (a \circ b)H$, for all $a, b \in G.$

Unit GTA4 Cosets and Lagrange's Theorem

8. Reminder: normal subgroups

A subgroup H of a group G is **normal** in G if, for each $g \in G$,

$$gH = Hg.$$

H ◁ G

9. Cosets of A_3 in S_3

o	e	(1 2 3)	(1 3 2)	(2 3)	(1 3)	(1 2)
e	e	(1 2 3)	(1 3 2)	(2 3)	(1 3)	(1 2)
(1 2 3)	(1 2 3)	(1 3 2)	e	(1 2)	(2 3)	(1 3)
(1 3 2)	(1 3 2)	e	(1 2 3)	(1 3)	(1 2)	(2 3)
(2 3)	(2 3)	(1 3)	(1 2)	e	(1 2 3)	(1 3 2)
(1 3)	(1 3)	(1 2)	(2 3)	(1 3 2)	e	(1 2 3)
(1 2)	(1 2)	(2 3)	(1 3)	(1 2 3)	(1 3 2)	e

$A_3 = \{\ e,\ (1\,2\,3),\ (1\,3\,2)\}$
$(1\,2)A_3 = \{(1\,2),\ (1\,3),\ (2\,3)\ \}$

	A_3	$(1\,2)A_3$
A_3	A_3	$(1\,2)A_3$
$(1\,2)A_3$	$(1\,2)A_3$	A_3

all subgroups of index 2 are normal

$+_2$	0	1
0	0	1
1	1	0

$\mathbb{Z}_2$:

10. Cosets of $H = \{e, (1\,2)\}$ in S_3

H is NOT normal in S_3

$H=\{e,(1\,2)\}$, $(2\,3)H=\{(2\,3),(1\,3\,2)\}$, $H(2\,3)=\{(2\,3),(1\,2\,3)\}$

different sets

$H\cdot(2\,3)H=\{(2\,3),(1\,2\,3),(1\,3\,2),(1\,3)\}$ – NOT a coset.

11. Cosets of $\{0,3\}$ in $(\mathbb{Z}_6, +_6)$

$+_6$	0	3	1	4	2	5
0	0	3	1	4	2	5
3	3	0	4	1	5	2
1	1	4	2	5	3	0
4	4	1	5	2	0	3
2	2	5	3	0	4	1
5	5	2	0	3	1	4

$H = \{0,3\}$
$1+H = \{1,4\}$
$2+H = \{2,5\}$

	H	1+H	2+H
H	H	1+H	2+H
1+H	1+H	2+H	H
2+H	2+H	H	1+H

all subgroups of an Abelian group are normal

$+_3$	0	1	2
0	0	1	2
1	1	2	0
2	2	0	1

$\mathbb{Z}_3$:

12. An important result

Theorem 5.1 Let N be a normal subgroup of G.

Then, for all $a, b \in G$,

(proof in text)

multiplicative $aN \cdot bN = (a \circ b)N$ (Frame 9)

additive $(a + N) + (b + N) = (a + b) + N$ (Frame 11)

13. Quotient groups

Theorem 5.2 Let N be a normal subgroup of G.

Then the set of cosets of N in G, with the operation $\cdot$, is a group, called the **quotient group**, denoted by G/N.

(proof in text)

(group table blocks)

(blocks are cosets)

- identity in G/N is N
- inverse of gN is $g^{-1}N$

14. The order of the quotient group G/N

$S_3/A_3 \cong \mathbb{Z}_2$ and $|S_3|/|A_3| = \frac{6}{3} = 2$ (Frame 9)

$\mathbb{Z}_6/\{0,3\} \cong \mathbb{Z}_3$ and $|\mathbb{Z}_6|/|\{0,3\}| = \frac{6}{2} = 3$ (Frame 11)

If G is a finite group, then $|G/N| = |G|/|N|$.

(number of cosets = **index** of N in G)

15. Exercise 5.5

(a) Explain why the subgroup $\{0,4\}$ is normal in $\mathbb{Z}_8$.

(b) For the quotient group $\mathbb{Z}_8/\{0,4\}$,

construct the Cayley table;

write down the identity element;

write down the inverse of each element.

(c) To which familiar group is $\mathbb{Z}_8/\{0,4\}$ isomorphic?

Post-audio exercises

In the audio section we showed that if a group G has a normal subgroup N, then the set of cosets of N in G itself forms a group G/N with the operation of set composition. If G is a finite group, the order of G/N is equal to the order of G divided by the order of N. For a finite group G of small order, we can find a familiar group (from the course) isomorphic to G/N by using the following strategy.

Strategy 5.1   To find a group isomorphic to a *finite* quotient group G/N, where N is a normal subgroup of a group $(G, \circ)$.

1. Calculate $|G/N| = |G|/|N| = k$, say.

2. Determine the k cosets aN by choosing different elements a until all the elements of G are assigned to cosets.

3. Construct the $k \times k$ group table of G/N by composing each pair of cosets, using the rule

    $$aN \cdot bN = (a \circ b)N - \text{coset containing } a \circ b.$$

4. By inspection of the group table, identify a familiar group (from the course) isomorphic to G/N.

In the group table, avoid writing a given coset in more than one way. For example, if $cN = dN$, then denote every occurrence of this coset by cN (or every occurrence by dN).

The following exercises give you an opportunity to use Strategy 5.1.

Exercise 5.6   Consider the group $\mathbb{Z}_{12}$ and its subgroup $H = \{0, 6\}$.

(a) Give a reason why $\{0, 6\}$ is normal in $\mathbb{Z}_{12}$.

(b) Construct the group table for the quotient group $\mathbb{Z}_{12}/\{0, 6\}$.

(c) To which familiar group is $\mathbb{Z}_{12}/\{0, 6\}$ isomorphic?

Exercise 5.7   Let G be the group defined by the group table on page 14.

For each of the following normal subgroups of G:

$$N = \{e, a, b, c, d, f, g, h, i, j, k, l\}, \quad N = \{e, a, b, c\},$$

(a) write down the cosets of N in G;

(b) construct the group table for the quotient group G/N;

(c) write down the inverse of each element of G/N;

(d) write down a familiar group isomorphic to G/N.

5.2 Proofs

We now supply the proofs of the theorems introduced in the audio section. First we prove Theorem 5.1, which states that set composition is a closed binary operation on the set of cosets whenever the subgroup is normal (it also provides a rule for the operation). The proof is not long, but it involves some quite intricate group theory. If you find the proof heavy going, then accept the result and move on; it is understanding the result itself and how to use it that matters.

Theorem 5.1 Let N be a normal subgroup of a group G. Then, for all $a, b \in G$,

$$aN \cdot bN = (ab)N.$$

As we mentioned earlier, we shall often use the letter N to denote a normal subgroup. Also, for ease of notation, we omit the group operation $\circ$ throughout this section.

If you are short of time, omit this proof.

Proof To show that the two sets $aN \cdot bN$ and $(ab)N$ are equal, we show first that $aN \cdot bN \subseteq (ab)N$ and then that $(ab)N \subseteq aN \cdot bN$.

(a) $aN \cdot bN \subseteq (ab)N$

A general element of the set $aN \cdot bN$ has the form

$$(an_1)(bn_2), \quad \text{where } n_1, n_2 \in N.$$

We show that this element can be written in the form $(ab)n$, for some element $n \in N$.

Now

$$(an_1)(bn_2) = a(n_1 b)n_2 \quad \text{(associativity).} \tag{5.1}$$

We know that $n_1 b$ is an element of the coset Nb and $Nb = bN$ (since N is a normal subgroup), so $n_1 b$ belongs to the coset bN. It follows that there is some element n_3, say, of N such that

$$n_1 b = bn_3.$$

Substituting this expression for $n_1 b$ in equation (5.1), we get

$$(an_1)(bn_2) = a(bn_3)n_2$$
$$= (ab)(n_3 n_2) \quad \text{(associativity).}$$

Since N is a subgroup, it follows from the closure axiom that $n_3 n_2 \in N$, so there is some element n of N such that $n_3 n_2 = n$; we conclude that

$$(an_1)(bn_2) = (ab)n$$

and so $(an_1)(bn_2) \in (ab)N$.

Hence

$$aN \cdot bN \subseteq (ab)N. \tag{5.2}$$

(b) $(ab)N \subseteq aN \cdot bN$

A general element of the set $(ab)N$ has the form

$$(ab)n, \quad \text{where } n \in N.$$

We show that this element is in the set $aN \cdot bN$.

Now

$$(ab)n = a(bn) \quad \text{(associativity).}$$

Since $a \in aN$ and $bn \in bN$, this shows that

$$(ab)n \in aN \cdot bN.$$

Hence

$$(ab)N \subseteq aN \cdot bN. \tag{5.3}$$

It follows from statements (5.2) and (5.3) that

$$aN \cdot bN = (ab)N. \quad \blacksquare$$

Next we prove our main result of this section.

Theorem 5.2   Let N be a normal subgroup of a group G. Then the set of cosets of N in G, with the operation of set composition, is a group.

Proof   We show that $(G/N, \cdot)$ satisfies the four group axioms.

If you are short of time, omit this proof.

G1 CLOSURE   Let aN, bN be any cosets of N in G. Then, by Theorem 5.1, we know that

$$aN \cdot bN = (ab)N,$$

which is also a coset of N in G. It follows that the set of cosets is closed under set composition.

G2 IDENTITY   Let e be the identity in G; then, for all $a \in G$,

$$ae = a = ea.$$

Thus, for all $aN \in G/N$,

$$aN \cdot eN = (ae)N = aN,$$
$$eN \cdot aN = (ea)N = aN,$$

so eN is an identity in G/N.

G3 INVERSES   Let a be any element of G; then there exists an inverse $a^{-1} \in G$ such that

$$aa^{-1} = e = a^{-1}a.$$

Thus, for each coset $aN \in G/N$,

$$aN \cdot a^{-1}N = (aa^{-1})N = eN = N,$$
$$a^{-1}N \cdot aN = (a^{-1}a)N = eN = N,$$

so $a^{-1}N$ is an inverse of aN in G/N.

G4 ASSOCIATIVITY   Let aN, bN and cN be any cosets of N in G. Then, by Theorem 5.1,

$$aN \cdot (bN \cdot cN) = aN \cdot (bc)N = (a(bc))N,$$
$$(aN \cdot bN) \cdot cN = (ab)N \cdot cN = ((ab)c)N.$$

These two expressions are the same because of the associativity of G. Therefore set composition is associative on G/N.

Hence G/N, the set cosets of N in G, with the operation of set composition, is a group.   ∎

Further exercises

Exercise 5.8 Let $(Q, \circ)$ be the group with the following group table.

$\circ$	e	i	j	k	w	x	y	z
e	e	i	j	k	w	x	y	z
i	i	w	k	y	x	e	z	j
j	j	z	w	i	y	k	e	x
k	k	j	x	w	z	y	i	e
w	w	x	y	z	e	i	j	k
x	x	e	z	j	i	w	k	y
y	y	k	e	x	j	z	w	i
z	z	y	i	e	k	j	x	w

This is the same group as in Exercises 1.8 and 4.8.

Determine the following set composites in the group $(Q, \circ)$.

(a) $\{e, w\} \cdot \{i, x\}$ (b) $\{j, y\} \cdot \{j, y\}$ (c) $\{k, z\} \cdot \{i, x\}$

Exercise 5.9 Let H be the subgroup $\{e, w\}$ of the group $(Q, \circ)$ in Exercise 5.8.

Determine the following set composites.

(a) $iH \cdot jH$ (b) $jH \cdot jH$ (c) $(iH) \cdot (jH \cdot kH)$

Exercise 5.10 Let H be the normal subgroup $\{e, w\}$ of the group $(Q, \circ)$ in Exercise 5.8.

(a) Construct the group table for Q/H.

(b) To which familiar group is Q/H isomorphic?

Exercise 5.11 Consider the group $(\mathbb{Z}_{13}^*, \times_{13})$ and its subgroup $H = \{1, 12\}$.

(a) Give a reason why H is a normal subgroup.

(b) Construct the group table for $\mathbb{Z}_{13}^*/H$.

(c) Show that $\mathbb{Z}_{13}^*/H$ is cyclic and therefore isomorphic to $\mathbb{Z}_6$.

You found the cosets of H in $\mathbb{Z}_{13}^*$ in Exercise 1.11.

6   Quotient groups of infinite groups

After working through this section, you should be able to:

(a) describe the quotient group $\mathbb{Z}/n\mathbb{Z}$ for $n \geq 2$;

(b) determine some quotient groups of infinite order, including the quotient group $\mathbb{R}/\mathbb{Z}$.

You may find this section, especially Subsection 6.3, more difficult than the rest of this unit. You may find it helpful to reread it after you have studied quotient groups further in Unit GTB2.

6.1 Quotient groups of $\mathbb{Z}$

In the examples of quotient groups G/N that we have seen so far, G has always been a finite group and the order of G/N has been small, so we have been able to construct the group table for G/N. From the group table, we have been able to identify a familiar group isomorphic to G/N.

We now become more ambitious and investigate quotient groups G/N where the group G is infinite. We begin by looking at examples where the quotient group is finite.

Consider the group $\mathbb{Z}$ of integers under addition. The group $\mathbb{Z}$ is cyclic, so all the subgroups of $\mathbb{Z}$ are cyclic—they are the trivial group $\{0\}$, the whole group $\mathbb{Z}$ and the subgroups

See Unit GTA2, Theorem 4.4.

$$n\mathbb{Z} = \langle n \rangle = \{\ldots, -3n, -2n, -n, 0, n, 2n, 3n, \ldots\}, \quad \text{for each } n \geq 2.$$

As $\mathbb{Z}$ is an *Abelian* group, *all* its subgroups are normal. Consequently, the quotient group $\mathbb{Z}/n\mathbb{Z}$ exists for each value of $n \geq 2$.

Theorem 4.2

Example 6.1   Determine the set of cosets of $5\mathbb{Z}$ in $\mathbb{Z}$ and hence state a familiar group from the course to which $\mathbb{Z}/5\mathbb{Z}$ is isomorphic.

Solution   There are precisely five distinct cosets of $5\mathbb{Z}$ in $\mathbb{Z}$:

$$0 + 5\mathbb{Z} = \{\ldots, -15, -10, -5, 0, 5, 10, 15, \ldots\},$$

$0 + 5\mathbb{Z} = 5\mathbb{Z}$

$$1 + 5\mathbb{Z} = \{\ldots, -14, -9, -4, 1, 6, 11, 16, \ldots\},$$
$$2 + 5\mathbb{Z} = \{\ldots, -13, -8, -3, 2, 7, 12, 17, \ldots\},$$
$$3 + 5\mathbb{Z} = \{\ldots, -12, -7, -2, 3, 8, 13, 18, \ldots\},$$
$$4 + 5\mathbb{Z} = \{\ldots, -11, -6, -1, 4, 9, 14, 19, \ldots\}.$$

These five cosets between them contain every integer, so they form a partition of $\mathbb{Z}$. Any other coset must be equal to one of these. For example,

$$5 + 5\mathbb{Z} = \{\ldots, -10, -5, 0, 5, 10, 15, 20, \ldots\} = 0 + 5\mathbb{Z},$$

$5 \in 0 + 5\mathbb{Z}$, so $5 + 5\mathbb{Z} = 0 + 5\mathbb{Z}$,

$$6 + 5\mathbb{Z} = \{\ldots, -9, -4, 1, 6, 11, 16, 21, \ldots\} = 1 + 5\mathbb{Z},$$

$6 \in 1 + 5\mathbb{Z}$, so $6 + 5\mathbb{Z} = 1 + 5\mathbb{Z}$,

$$-8 + 5\mathbb{Z} = \{\ldots, -23, -18, -13, -8, -3, 2, 7, \ldots\} = 2 + 5\mathbb{Z}.$$

$-8 \in 2 + 5\mathbb{Z}$, so $-8 + 5\mathbb{Z} = 2 + 5$

So the underlying set of the group $\mathbb{Z}/5\mathbb{Z}$ has just five elements, namely the above five cosets. Thus $\mathbb{Z}/5\mathbb{Z}$ is a group of order 5, so it is cyclic and hence isomorphic to $\mathbb{Z}_5$.   ∎

Let us find an isomorphism between $\mathbb{Z}/5\mathbb{Z}$ and $\mathbb{Z}_5$. To help us, we first construct the group table for $\mathbb{Z}/5\mathbb{Z}$.

The rule for composition of cosets in an additive group is

$$(a + N) + (b + N) = (a + b) + N.$$

Here $N = 5\mathbb{Z}$, so we have

$$(a + 5\mathbb{Z}) + (b + 5\mathbb{Z}) = (a + b) + 5\mathbb{Z};$$

for example,

$$(1 + 5\mathbb{Z}) + (2 + 5\mathbb{Z}) = 3 + 5\mathbb{Z},$$
$$(4 + 5\mathbb{Z}) + (2 + 5\mathbb{Z}) = 6 + 5\mathbb{Z} = 1 + 5\mathbb{Z}.$$

All the other calculations are carried out similarly and we obtain the following group table.

$+$	$0 + 5\mathbb{Z}$	$1 + 5\mathbb{Z}$	$2 + 5\mathbb{Z}$	$3 + 5\mathbb{Z}$	$4 + 5\mathbb{Z}$
$0 + 5\mathbb{Z}$	$0 + 5\mathbb{Z}$	$1 + 5\mathbb{Z}$	$2 + 5\mathbb{Z}$	$3 + 5\mathbb{Z}$	$4 + 5\mathbb{Z}$
$1 + 5\mathbb{Z}$	$1 + 5\mathbb{Z}$	$2 + 5\mathbb{Z}$	$3 + 5\mathbb{Z}$	$4 + 5\mathbb{Z}$	$0 + 5\mathbb{Z}$
$2 + 5\mathbb{Z}$	$2 + 5\mathbb{Z}$	$3 + 5\mathbb{Z}$	$4 + 5\mathbb{Z}$	$0 + 5\mathbb{Z}$	$1 + 5\mathbb{Z}$
$3 + 5\mathbb{Z}$	$3 + 5\mathbb{Z}$	$4 + 5\mathbb{Z}$	$0 + 5\mathbb{Z}$	$1 + 5\mathbb{Z}$	$2 + 5\mathbb{Z}$
$4 + 5\mathbb{Z}$	$4 + 5\mathbb{Z}$	$0 + 5\mathbb{Z}$	$1 + 5\mathbb{Z}$	$2 + 5\mathbb{Z}$	$3 + 5\mathbb{Z}$

If we drop all the references to '$+5\mathbb{Z}$' in this table, then we obtain the familiar group table for $\mathbb{Z}_5$.

	0	1	2	3	4
0	0	1	2	3	4
1	1	2	3	4	0
2	2	3	4	0	1
3	3	4	0	1	2
4	4	0	1	2	3

So an isomorphism between the two groups is the function

$$\phi : \mathbb{Z}/5\mathbb{Z} \longrightarrow \mathbb{Z}_5$$
$$n + 5\mathbb{Z} \longmapsto n, \quad \text{for } n = 0, 1, 2, 3, 4.$$

The five cosets of $\mathbb{Z}/5\mathbb{Z}$ are precisely the equivalence classes for congruence modulo 5: two integers belong to the same coset of $5\mathbb{Z}$ in $\mathbb{Z}$ if and only if they are congruent modulo 5.

The function ϕ renames the elements in the table for $\mathbb{Z}/5\mathbb{Z}$ to give the table for $\mathbb{Z}_5$.

To see whether the result $\mathbb{Z}/5\mathbb{Z} \cong \mathbb{Z}_5$ is a consequence of the fact that 5 is a prime, let us repeat this exercise for a non-prime integer.

Exercise 6.1   Determine the set of cosets of $6\mathbb{Z}$ in $\mathbb{Z}$ and construct the group table for the group $\mathbb{Z}/6\mathbb{Z}$. Decide whether this group is cyclic and state a familiar group to which $\mathbb{Z}/6\mathbb{Z}$ is isomorphic.

All the evidence suggests that $\mathbb{Z}/n\mathbb{Z}$ is a cyclic group isomorphic to $\mathbb{Z}_n$, for any integer $n \geq 2$. Certainly, $\mathbb{Z}/n\mathbb{Z}$ has order n because we can list its n elements; they are the cosets

$$0 + n\mathbb{Z}, \quad 1 + n\mathbb{Z}, \quad 2 + n\mathbb{Z}, \quad \ldots, \quad (n-1) + n\mathbb{Z}.$$

But why must $\mathbb{Z}/n\mathbb{Z}$ be cyclic? Remember that a group is cyclic if one of its elements generates the whole group.

Exercise 6.2   For each of the following quotient groups, find an element which generates the whole group.

(a) $\mathbb{Z}/5\mathbb{Z}$ (b) $\mathbb{Z}/6\mathbb{Z}$

Which element would you expect to generate $\mathbb{Z}/n\mathbb{Z}$, for $n \geq 2$?

To confirm that the coset $1 + n\mathbb{Z}$ generates $\mathbb{Z}/n\mathbb{Z}$, we consider the successive multiples of $1 + n\mathbb{Z}$:

$$1 + n\mathbb{Z},$$
$$2(1 + n\mathbb{Z}) = (1 + n\mathbb{Z}) + (1 + n\mathbb{Z}) = 2 + n\mathbb{Z},$$
$$3(1 + n\mathbb{Z}) = (2 + n\mathbb{Z}) + (1 + n\mathbb{Z}) = 3 + n\mathbb{Z},$$
$$4(1 + n\mathbb{Z}) = (3 + n\mathbb{Z}) + (1 + n\mathbb{Z}) = 4 + n\mathbb{Z}$$

and, in general, for $m \in \mathbb{Z}$,

$$m(1 + n\mathbb{Z}) = m + n\mathbb{Z}.$$

Thus every coset of $n\mathbb{Z}$ in $\mathbb{Z}$ is a multiple of the coset $1 + n\mathbb{Z}$; that is, $1 + n\mathbb{Z}$ generates $\mathbb{Z}/n\mathbb{Z}$.

We summarise our results in the following theorem.

> **Theorem 6.1** For each integer $n \geq 2$, the quotient group $\mathbb{Z}/n\mathbb{Z}$ is a cyclic group of order n generated by $1 + n\mathbb{Z}$ and so
>
> $$\mathbb{Z}/n\mathbb{Z} \cong \mathbb{Z}_n.$$

6.2 The quotient group $\mathbb{R}/\mathbb{Z}$

We now consider an example in which the quotient group has infinite order. In this case we cannot construct a group table, so we need some other way of describing the quotient group. The first task is to describe the underlying set of the quotient group: we need a workable description for the infinite set of cosets.

This section is devoted to one group—the quotient group $\mathbb{R}/\mathbb{Z}$.

Since addition is commutative, the group $\mathbb{R}$ is Abelian. All subgroups of an Abelian group are normal, so the subgroup $\mathbb{Z}$ is normal. The quotient group $\mathbb{R}/\mathbb{Z}$ therefore exists. But is it a group that we have already met?

See Theorem 4.2.

The elements of $\mathbb{R}/\mathbb{Z}$ are the distinct cosets of $\mathbb{Z}$ in $\mathbb{R}$. We begin by determining some of these cosets. For any real number $r \in \mathbb{R}$, the coset to which r belongs is

$$r + \mathbb{Z} = \{\ldots, r-2, r-1, r, r+1, r+2, \ldots\},$$

since this is the set obtained by adding each of the integers (positive, negative and 0) to r. For example,

$$1 + \mathbb{Z} = \{\ldots, -3, -2, -1, 0, 1, 2, 3, \ldots\} = \mathbb{Z}$$

and

$$1.2 + \mathbb{Z} = \{\ldots, -2.8, -1.8, -0.8, 0.2, 1.2, 2.2, 3.2, \ldots\}.$$

> **Exercise 6.3** Consider the following list of five cosets of $\mathbb{Z}$ in $\mathbb{R}$:
>
> $$3.7 + \mathbb{Z}, \quad 0.1 + \mathbb{Z}, \quad -1.3 + \mathbb{Z}, \quad 39.1 + \mathbb{Z}, \quad -4.9 + \mathbb{Z}.$$
>
> (a) How many different cosets are there in this list?
>
> (b) For each coset, determine the smallest positive real number that belongs to it.

Let us consider the coset $1.2 + \mathbb{Z}$. It consists of all the real numbers of the form

$$(\text{an integer}) + 0.2.$$

For example,

$$1.2 = 1 + 0.2,$$
$$-2.8 = -3 + 0.2.$$

We say that such numbers have *fractional part* 0.2. Thus this particular coset consists of all the real numbers with fractional part 0.2. We represent the coset pictorially as follows.

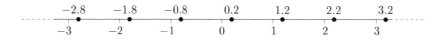

The fractional part of a real number x is $x - [x]$, where $[x]$ is the largest integer not exceeding x; for example, the fractional part of 1.2 is

$$1.2 - 1 = 0.2$$

and the fractional part of -2.8 is

$$-2.8 - (-3) = -2.8 + 3 = 0.2$$

The elements of the coset are evenly spaced along the real line at unit distance apart. The element of the coset which lies in the interval $[0, 1)$, namely 0.2, is the fractional part shared by all the elements of the coset.

In general, the elements of a coset all have the same fractional part. Moreover, this fractional part is equal to the unique element of the coset that lies in the interval $[0, 1)$. In this sense, each coset corresponds to the unique real number in $[0, 1)$ that lies in the coset. Conversely, each real number $r \in [0, 1)$ corresponds to the coset of $\mathbb{Z}$ in $\mathbb{R}$ which contains r, namely $r + \mathbb{Z}$. The set of cosets of $\mathbb{Z}$ in $\mathbb{R}$ can therefore be described as

$$\{r + \mathbb{Z} : r \in [0, 1)\}.$$

Exercise 6.4 For each of the following cosets of $\mathbb{Z}$ in $\mathbb{R}$, find the corresponding unique fractional part and hence write the coset in the form $r + \mathbb{Z}$, where $r \in [0, 1)$:

$$3.1 + \mathbb{Z}, \quad -3.1 + \mathbb{Z}, \quad \sqrt{2} + \mathbb{Z}.$$

We now have a description of the set of cosets of $\mathbb{Z}$ in $\mathbb{R}$, so we know the *elements* of the quotient group. Next we turn to the binary operation of the quotient group. As usual we combine cosets using set composition. Since the binary operation on $\mathbb{R}$ is addition, it follows from Theorem 5.1 that the formula for the set composition is

$$(x + \mathbb{Z}) + (y + \mathbb{Z}) = (x + y) + \mathbb{Z}.$$

For example,

$$(0.25 + \mathbb{Z}) + (0.7 + \mathbb{Z}) = (0.25 + 0.7) + \mathbb{Z} = 0.95 + \mathbb{Z}$$

and

$$(0.3 + \mathbb{Z}) + (0.96 + \mathbb{Z}) = 1.26 + \mathbb{Z} = 0.26 + \mathbb{Z}.$$

Notice that, in the second example above, addition of the two fractional parts on the left gives 1.26, which is not a fractional part, so if we wish to work exclusively with fractional parts, then we must subtract 1 in this case to obtain the fractional part 0.26.

The rule for combining the cosets $x + \mathbb{Z}$ and $y + \mathbb{Z}$, where $x \in [0, 1)$ and $y \in [0, 1)$, is

$$(x + \mathbb{Z}) + (y + \mathbb{Z}) = \begin{cases} (x + y) + \mathbb{Z}, & \text{if } x + y < 1, \\ (x + y - 1) + \mathbb{Z}, & \text{if } x + y \geq 1. \end{cases}$$

$0 \leq x + y < 2$, so $x + y$ belongs to either $[0, 1)$ or $[1, 2)$.

The above rule for composing cosets may seem familiar. If we ignore the '$+\mathbb{Z}$' part and concentrate on the fractional parts, it is just addition modulo 1 on the set $[0, 1)$, since this can be defined by

$$x +_1 y = \begin{cases} x + y, & \text{if } x + y < 1, \\ x + y - 1, & \text{if } x + y \geq 1. \end{cases}$$

This is similar to addition modulo 2π, discussed in Unit I3, Subsection 4.2.

Thus the rule for composition of elements of $\mathbb{R}/\mathbb{Z}$ can be given more succinctly as

$$(x + \mathbb{Z}) + (y + \mathbb{Z}) = (x +_1 y) + \mathbb{Z}.$$

Exercise 6.5 Determine the following cosets of $\mathbb{Z}$ in $\mathbb{R}$.

(a) $(0.3 + \mathbb{Z}) + (0.42 + \mathbb{Z})$

(b) $(0.63 + \mathbb{Z}) + (0.63 + \mathbb{Z})$

(c) $(0.63 + \mathbb{Z}) + (0.37 + \mathbb{Z})$

We now identify a group isomorphic to $\mathbb{R}/\mathbb{Z}$. We have seen that every element of $\mathbb{R}/\mathbb{Z}$ corresponds to an element of $[0, 1)$, and vice-versa. We have also seen that composition of elements in $\mathbb{R}/\mathbb{Z}$ corresponds to addition modulo 1 of elements of $[0, 1)$. This suggests that

$$(\mathbb{R}/\mathbb{Z}, +) \cong ([0, 1), +_1).$$

But is $([0, 1), +_1)$ a group? You are asked to show that it is in the next exercise.

Exercise 6.6 Show that $([0, 1), +_1)$ is a group.

Hint: To prove that $+_1$ is associative, consider arbitrary x, y and z in $[0, 1)$. Show that $(x +_1 y) +_1 z$ is equal to $x +_1 (y +_1 z)$ plus some integer, and then explain why this integer is zero.

The proof that $+_1$ is associative is a little tricky. If you find yourself struggling with this part of the exercise just read through the solution.

Now consider the function

$$\phi : \mathbb{R}/\mathbb{Z} \longrightarrow [0, 1)$$
$$r + \mathbb{Z} \longmapsto \text{fractional part of } r.$$

From the discussion above we know that this is a one-one and onto function. Also, if $a + \mathbb{Z}$ and $b + \mathbb{Z}$ are two cosets of $\mathbb{Z}$ in $\mathbb{R}$ (expressed so that both a and b lie in $[0, 1)$), then

$$\phi((a + \mathbb{Z}) + (b + \mathbb{Z})) = \phi((a +_1 b) + \mathbb{Z})$$
$$= a +_1 b$$
$$= \phi(a + \mathbb{Z}) +_1 \phi(b + \mathbb{Z}).$$

This shows that ϕ is an isomorphism from $\mathbb{R}/\mathbb{Z}$ to $([0, 1), +_1)$, giving us the following result.

Theorem 6.2 The quotient group $\mathbb{R}/\mathbb{Z}$ is isomorphic to $([0, 1), +_1)$, the group of real numbers in the interval $[0, 1)$, under addition modulo 1.

The next exercise presents a similar quotient group for you to identify. To do this, and to identify other similar quotient groups, you need to be familiar with a family of groups that generalise the group in Exercise 6.6 as follows. For any positive real number m, we define addition modulo m on the interval $[0, m)$ by

$$x +_m y = \begin{cases} x + y, & \text{if } x + y < m, \\ x + y - m, & \text{if } x + y \geq m. \end{cases}$$

$0 \leq x + y < 2m$, so $x + y$ belongs to either $[0, m)$ or $[m, 2m)$.

Then $([0, m), +_m)$ is a group; the proof is similar to the solution to Exercise 6.6, but we omit the details.

Exercise 6.7 This exercise is about the quotient group $\mathbb{R}/3\mathbb{Z}$.

(a) Write down three positive numbers and three negative numbers in each of the following cosets:

$$4.3 + 3\mathbb{Z}, \quad 0.75 + 3\mathbb{Z}, \quad -12.9 + 3\mathbb{Z}.$$

Recall that $3\mathbb{Z}$ is the subgroup of $\mathbb{Z}$ (and hence also of $\mathbb{R}$) consisting of all integer multiple of 3 (see Unit GTA2, Exercise 1.2(a)).

(b) Give a simple description of the set of all the cosets of $3\mathbb{Z}$ in $\mathbb{R}$ in the form
$$\{r + 3\mathbb{Z} : r \in ?\}.$$

(c) Determine the following cosets in the quotient group $\mathbb{R}/3\mathbb{Z}$:
$$(1 + 3\mathbb{Z}) + (2.3 + 3\mathbb{Z}), \quad (2.5 + 3\mathbb{Z}) + (0 + 3\mathbb{Z}),$$
$$(1.93 + 3\mathbb{Z}) + (0.75 + 3\mathbb{Z}).$$

(d) Write down a simple description of the binary operation on the set of cosets in the quotient group $\mathbb{R}/3\mathbb{Z}$.

(e) Write down a result, similar to that of Theorem 6.2, giving a group that is isomorphic to $\mathbb{R}/3\mathbb{Z}$.

6.3 An example from $\mathbb{R}^2$

Earlier in this block you met the following example of an infinite group: the underlying set is

See Unit GTA2, Example 1.5.

$$X = \{(a, b) \in \mathbb{R}^2 : a \neq 0\}$$

This is a subset of $\mathbb{R}^2$.

and the binary operation $*$ is defined on X by

$$(a, b) * (c, d) = (ac, ad + b).$$

We discovered that the identity element of this group is $(1, 0)$ and the inverse of the element (a, b) is $(1/a, -b/a)$. We also showed that

$$A = \{(1, b) : b \in \mathbb{R}\}$$

is a subgroup of $(X, *)$.

In this subsection we explore the quotient group X/A. Our first task is to confirm that this quotient group exists by confirming that A is a normal subgroup of X. We begin by finding the left coset of A in X.

Exercise 6.8 Determine the following left cosets of A in X.

(a) $(2, 1)A$ (b) $(2, 3)A$ (c) $(-2, \pi)A$

The solution to Exercise 6.8 suggests that each left coset of A in X consists of all the elements of X with a particular number in the first coordinate position.

We can determine whether this is the case by finding a general left coset $(x, y)A$:

$$\begin{aligned}(x, y)A &= \{(x, y) * (1, b) : b \in \mathbb{R}\}\\ &= \{(x, bx + y) : b \in \mathbb{R}\}\\ &= \{(x, r) : r \in \mathbb{R}\}.\end{aligned}$$

Here $x \neq 0$, so as b ranges over all real values, $bx + y$ ranges over all real values.

So the left coset $(x, y)A$ does indeed consist of all the elements of X with first coordinate x. Let us check whether A is a normal subgroup of X by finding the *right* coset $A(x, y)$:

$$\begin{aligned}A(x, y) &= \{(1, b) * (x, y) : b \in \mathbb{R}\}\\ &= \{(x, y + b) : b \in \mathbb{R}\}\\ &= \{(x, r) : r \in \mathbb{R}\}\\ &= (x, y)A.\end{aligned}$$

Here, as b ranges over all real values, $y + b$ ranges over all real values.

We have found that $(x, y)A = A(x, y)$ for all $(x, y) \in X$, so A is a normal subgroup of X.

Hence the quotient group X/A exists. Its elements are the cosets of A in X, which we have seen are the sets of elements of X with a common first coordinate. The coset of A containing all the elements with first coordinate $x \in \mathbb{R}^*$ contains the particular element $(x, 0)$, so it can be denoted by

$$(x, 0)A$$

and thus

$$X/A = \{(x, 0)A : x \in \mathbb{R}^*\}.$$

Exercise 6.9 For each of the following elements of X, write down the coset to which it belongs in the form $(x, 0)A$, where $x \in \mathbb{R}^*$.

(a) $(17, -3)$ (b) $(-\sqrt{2}, \sqrt{2})$ (c) $(1, \frac{3}{4})$

Now that we have a workable description of the set X/A, we consider the binary operation on this set. It is given by

$$(x, 0)A \cdot (y, 0)A = ((x, 0) * (y, 0))A;$$

that is,

$$(x, 0)A \cdot (y, 0)A = (xy, 0)A.$$

Theorem 5.1 tells us that set composition on X/A is given by this formula.

Exercise 6.10

(a) Determine the following composites in the group X/A:

$$(3.2, 0)A \cdot (-3, 0)A, \quad (\tfrac{7}{3}, 0)A \cdot (-\tfrac{2}{5}, 0)A.$$

(b) To which coset $(x, 0)A$ does the composite $(3, -2) * (-2, 3)$ belong?

We now have a complete description of the quotient group X/A: we have an underlying set, together with a rule for the binary operation. Can we find a familiar group isomorphic to X/A? We know that each coset corresponds to a non-zero real number—the number that is the first coordinate of each element of the coset. We also know, from the binary operation on X/A, that cosets are combined by multiplying these non-zero real numbers. This suggests that

$$X/A \cong \mathbb{R}^*$$

and that the function

$$\phi : X/A \longrightarrow \mathbb{R}^*$$
$$(x, 0)A \longmapsto x,$$

is an isomorphism. The function ϕ is certainly one-one and onto, and it preserves composites since, for all cosets $(x, 0)A$ and $(y, 0)A$, we have

$$\phi((x, 0)A \cdot (y, 0)A) = \phi((xy, 0)A)$$
$$= xy$$
$$= \phi((x, 0)A)\phi((y, 0)A).$$

Thus ϕ is an isomorphism and X/A is indeed isomorphic to $\mathbb{R}^*$.

Further exercises

Exercise 6.11   Construct the group table for the group $\mathbb{Z}/4\mathbb{Z}$. Verify that this group is cyclic by showing that $1 + 4\mathbb{Z}$ is a generator.

Exercise 6.12   Consider the set

$$X = \{(a, b) \in \mathbb{R}^2 : a > 0, b > 0\}.$$

A binary operation $*$ is defined on X by

$$(a, b) * (c, d) = (ac, bd).$$

(a) Prove that $(X, *)$ is an Abelian group.

(b) Prove that the subset

$$A = \{(a, a) \in \mathbb{R}^2 : a > 0\}$$

is a subgroup of X. Is A a normal subgroup of X?

(c) Show that the coset $(1, 2)A$ consists of all the points of X that lie on the line $y = 2x$.

(d) Describe geometrically (as in part (c)) the cosets

$$(\tfrac{1}{2}, \tfrac{1}{3})A \quad \text{and} \quad (3, 5)A.$$

(e) Explain why the set of all cosets of A in X is

$$\{(1, b)A : b > 0\}.$$

(f) Complete the following formula for the binary operation on X/A:

$$(1, x)A \cdot (1, y)A = (1, ?)A.$$

(g) To which familiar group is X/A isomorphic?

Solutions to the exercises

1.1 (a) For $H = \{e, s\}$,

$$
\begin{aligned}
eH &= \{e \circ e, e \circ s\} &&= \{e, s\}, \\
aH &= \{a \circ e, a \circ s\} &&= \{a, t\}, \\
bH &= \{b \circ e, b \circ s\} &&= \{b, u\}, \\
cH &= \{c \circ e, c \circ s\} &&= \{c, r\}, \\
rH &= \{r \circ e, r \circ s\} &&= \{r, c\}, \\
sH &= \{s \circ e, s \circ s\} &&= \{s, e\}, \\
tH &= \{t \circ e, t \circ s\} &&= \{t, a\}, \\
uH &= \{u \circ e, u \circ s\} &&= \{u, b\}.
\end{aligned}
$$

There are four distinct cosets:

$$
\begin{aligned}
eH &= sH = \{e, s\}, \\
aH &= tH = \{a, t\}, \\
bH &= uH = \{b, u\}, \\
cH &= rH = \{c, r\}.
\end{aligned}
$$

(b) For $H = \{e, b, r, t\}$,

$$
\begin{aligned}
eH &= \{e \circ e, e \circ b, e \circ r, e \circ t\} &&= \{e, b, r, t\}, \\
aH &= \{a \circ e, a \circ b, a \circ r, a \circ t\} &&= \{a, c, s, u\}, \\
bH &= \{b \circ e, b \circ b, b \circ r, b \circ t\} &&= \{b, e, t, r\}, \\
cH &= \{c \circ e, c \circ b, c \circ r, c \circ t\} &&= \{c, a, u, s\}, \\
rH &= \{r \circ e, r \circ b, r \circ r, r \circ t\} &&= \{r, t, e, b\}, \\
sH &= \{s \circ e, s \circ b, s \circ r, s \circ t\} &&= \{s, u, a, c\}, \\
tH &= \{t \circ e, t \circ b, t \circ r, t \circ t\} &&= \{t, r, b, e\}, \\
uH &= \{u \circ e, u \circ b, u \circ r, u \circ t\} &&= \{u, s, c, a\}.
\end{aligned}
$$

There are two distinct cosets:

$$
\begin{aligned}
eH &= bH = rH = tH = \{e, b, r, t\}, \\
aH &= cH = sH = uH = \{a, c, s, u\}.
\end{aligned}
$$

1.2 The group S_3 has order 6, so there are six potential cosets of $H = \{e, (1\ 2)\}$. However, we expect three distinct cosets.

$$
\begin{aligned}
eH &= \{e \circ e, e \circ (1\ 2)\} \\
&= \{e, (1\ 2)\}, \\
(1\ 2)H &= \{(1\ 2) \circ e, (1\ 2) \circ (1\ 2)\} \\
&= \{(1\ 2), e\}, \\
(1\ 3)H &= \{(1\ 3) \circ e, (1\ 3) \circ (1\ 2)\} \\
&= \{(1\ 3), (1\ 2\ 3)\}, \\
(2\ 3)H &= \{(2\ 3) \circ e, (2\ 3) \circ (1\ 2)\} \\
&= \{(2\ 3), (1\ 3\ 2)\}, \\
(1\ 2\ 3)H &= \{(1\ 2\ 3) \circ e, (1\ 2\ 3) \circ (1\ 2)\} \\
&= \{(1\ 2\ 3), (1\ 3)\}, \\
(1\ 3\ 2)H &= \{(1\ 3\ 2) \circ e, (1\ 3\ 2) \circ (1\ 2)\} \\
&= \{(1\ 3\ 2), (2\ 3)\}.
\end{aligned}
$$

There are 3 distinct cosets:

$$
\begin{aligned}
eH &= (1\ 2)H = \{e, (1\ 2)\}, \\
(1\ 3)H &= (1\ 2\ 3)H = \{(1\ 3), (1\ 2\ 3)\}, \\
(2\ 3)H &= (1\ 3\ 2)H = \{(2\ 3), (1\ 3\ 2)\}.
\end{aligned}
$$

1.3 By definition,

$$
gH = \{g \circ h : h \in H\}.
$$

If $g \in H$, then, for each $h \in H$,

$$
g \circ h \in H \quad \text{(closure in } H\text{)},
$$

so

$$
gH \subseteq H.
$$

The subgroup is finite, so it follows that $gH = H$, since $gH \subseteq H$ and the cosets H and gH have the same number of elements, by property 1.

Conversely, if $gH = H$, then each element of gH belongs to H. In particular, since $e \in H$, the element $g \circ e = g$ belongs to gH, so

$$
g \in H.
$$

1.4 (a) The subgroup $H = \{e, a, b, c\}$ is one coset.

All the cosets of H contain the same number of elements as H, by property 1.

It follows that there is just one other coset—namely, $\{r, s, t, u\}$.

So the partition is

$$
\{e, a, b, c\}, \quad \{r, s, t, u\}.
$$

(b) All the cosets of H contain the same number of elements as H, by property 1.

Thus each coset of the subgroup $\{e\}$ has just one element.

Hence the partition consists of eight cosets, each with one element:

$$
\{e\}, \quad \{a\}, \quad \{b\}, \quad \{c\}, \quad \{r\}, \quad \{s\}, \quad \{t\}, \quad \{u\}.
$$

(c) The subgroup $S(\square)$ is the whole group, so it is the only coset; thus the partition is just $S(\square)$.

1.5 We use Strategy 1.1 for partitioning a group into cosets.

We take the subgroup H as the first coset:

$$
H = \{e, (1\ 3)\}.
$$

This does not contain the element $(1\ 2)$, so we take $(1\ 2)H$ as the second coset:

$$
(1\ 2)H = \{(1\ 2) \circ e, (1\ 2) \circ (1\ 3)\} = \{(1\ 2), (1\ 3\ 2)\}.
$$

These two cosets do not contain the element $(2\ 3)$, so we take $(2\ 3)H$ as the third coset:

$$
(2\ 3)H = \{(2\ 3) \circ e, (2\ 3) \circ (1\ 3)\} = \{(2\ 3), (1\ 2\ 3)\}.
$$

Every element of S_3 has now been assigned to a coset, so the partition is

$$
\{e, (1\ 3)\}, \quad \{(1\ 2), (1\ 3\ 2)\}, \quad \{(2\ 3), (1\ 2\ 3)\}.
$$

1.6 (a) We take the subgroup H as the first coset:

$$
H = \{0, 2, 4, 6, 8\}.
$$

This does not contain the element 1, so we take $1 + H$ as the second coset:

$$
\begin{aligned}
1 + H &= 1 + \{0, 2, 4, 6, 8\} \\
&= \{1 +_{10} 0, 1 +_{10} 2, 1 +_{10} 4, 1 +_{10} 6, 1 +_{10} 8\} \\
&= \{1, 3, 5, 7, 9\}.
\end{aligned}
$$

Every element of $\mathbb{Z}_{10}$ has now been assigned to a coset, so the partition is

$$
\{0, 2, 4, 6, 8\}, \quad \{1, 3, 5, 7, 9\}.
$$

(b) We take the subgroup H as the first coset:
$$H = \{0, 5\}.$$
This does not contain the element 1, so we take $1 + H$ as the second coset:
$$\begin{aligned} 1 + H &= 1 + \{0, 5\} \\ &= \{1 +_{10} 0, 1 +_{10} 5\} \\ &= \{1, 6\}. \end{aligned}$$
These two cosets do not contain the element 2, so we take $2 + H$ as the third coset:
$$\begin{aligned} 2 + H &= 2 + \{0, 5\} \\ &= \{2 +_{10} 0, 2 +_{10} 5\} \\ &= \{2, 7\}. \end{aligned}$$
These three cosets do not contain the element 3, so we take $3 + H$ as the fourth coset:
$$\begin{aligned} 3 + H &= 3 + \{0, 5\} \\ &= \{3 +_{10} 0, 3 +_{10} 5\} \\ &= \{3, 8\}. \end{aligned}$$
These four cosets do not contain the element 4, so we take $4 + H$ as the fifth coset:
$$\begin{aligned} 4 + H &= 4 + \{0, 5\} \\ &= \{4 +_{10} 0, 4 +_{10} 5\} \\ &= \{4, 9\}. \end{aligned}$$
Every element of $\mathbb{Z}_{10}$ has now been assigned to a coset, so the partition is
$$\{0, 5\}, \quad \{1, 6\}, \quad \{2, 7\}, \quad \{3, 8\}, \quad \{4, 9\}.$$

1.7 (a) The distinct cosets of $4\mathbb{Z}$ in $\mathbb{Z}$ are:
$$\begin{aligned} 4\mathbb{Z} &= \{\ldots, -8, -4, 0, 4, 8, \ldots\}, \\ 1 + 4\mathbb{Z} &= \{\ldots, -7, -3, 1, 5, 9, \ldots\}, \\ 2 + 4\mathbb{Z} &= \{\ldots, -6, -2, 2, 6, 10, \ldots\}, \\ 3 + 4\mathbb{Z} &= \{\ldots, -5, -1, 3, 7, 11, \ldots\}. \end{aligned}$$
(b) The distinct cosets of $6\mathbb{Z}$ in $2\mathbb{Z}$ are:
$$\begin{aligned} 6\mathbb{Z} &= \{\ldots, -12, -6, 0, 6, 12, \ldots\}, \\ 2 + 6\mathbb{Z} &= \{\ldots, -10, -4, 2, 8, 14, \ldots\}, \\ 4 + 6\mathbb{Z} &= \{\ldots, -8, -2, 4, 10, 16, \ldots\}. \end{aligned}$$
Remark Here the whole group $2\mathbb{Z}$ consists of only the *even* integers, so we do not have cosets such as $1 + 6\mathbb{Z}$.

1.8 We use Strategy 1.1 to partition the group Q into cosets.

(a) We take the subgroup H as the first coset:
$$H = \{e, w\}.$$
This does not contain the element i, so we take iH as the second coset:
$$iH = \{i \circ e, i \circ w\} = \{i, x\}.$$
These two cosets do not contain the element j, so we take jH as the third coset:
$$jH = \{j \circ e, j \circ w\} = \{j, y\}.$$

These three cosets do not contain the element k, so we take kH as the fourth coset:
$$kH = \{k \circ e, k \circ w\} = \{k, z\}.$$
Every element of Q has now been assigned to a coset, so the partition is
$$\{e, w\}, \quad \{i, x\}, \quad \{j, y\}, \quad \{k, z\}.$$
(b) We take the subgroup H as the first coset:
$$H = \{e, i, w, x\}.$$
This does not contain the element j, so we take jH as the second coset:
$$jH = \{j \circ e, j \circ i, j \circ w, j \circ x\} = \{j, z, y, k\}.$$
Every element of Q has now been assigned to a coset, so the partition is
$$\{e, i, w, x\}, \quad \{j, z, y, k\}.$$

1.9 (a) The coset $(1\ 2)H$ consists of all the elements of S_4 of the form
$$(1\ 2) \circ h, \quad \text{where } h \in H.$$
Let h be any element of H, so $h(1) = 1$, and consider the effect of the composite $(1\ 2) \circ h$ on the symbol 1:
$$1 \xrightarrow{h} 1 \xrightarrow{(1\ 2)} 2.$$
So each element of the coset $(1\ 2)H$ maps 1 to 2.
The coset $(1\ 2)H$ contains six elements (because H has order 6) and there are just six elements of S_4 which map 1 to 2, namely
$$\begin{aligned} &(1\ 2), \quad (1\ 2)(3\ 4), \quad (1\ 2\ 3), \\ &(1\ 2\ 4), \quad (1\ 2\ 3\ 4), \quad (1\ 2\ 4\ 3). \end{aligned}$$
So these are the six elements of $(1\ 2)H$.

(b) Similarly, the coset $(1\ 3)H$ consists of the six elements of S_4 which map 1 to 3, namely
$$\begin{aligned} &(1\ 3), \quad (1\ 3)(2\ 4), \quad (1\ 3\ 2), \\ &(1\ 3\ 4), \quad (1\ 3\ 2\ 4), \quad (1\ 3\ 4\ 2), \end{aligned}$$
and $(1\ 4)H$ consists of the six elements of S_4 which map 1 to 4, namely
$$\begin{aligned} &(1\ 4), \quad (1\ 4)(2\ 3), \quad (1\ 4\ 2), \\ &(1\ 4\ 3), \quad (1\ 4\ 2\ 3), \quad (1\ 4\ 3\ 2). \end{aligned}$$

1.10 (a) We take the subgroup H as the first coset:
$$H = \{0, 9\}.$$
This does not contain the element 1, so we take $1 + H$ as the second coset:
$$1 + H = \{1 +_{18} 0, 1 +_{18} 9\} = \{1, 10\}.$$
These two cosets do not contain the element 2, so we take $2 + H$ as the third coset:
$$2 + H = \{2 +_{18} 0, 2 +_{18} 9\} = \{2, 11\}.$$
Continuing this process, using the elements 3, 4, 5, 6, 7 and 8 in turn, we obtain the cosets:
$$\begin{aligned} &\{3, 12\}, \quad \{4, 13\}, \quad \{5, 14\}, \\ &\{6, 15\}, \quad \{7, 16\}, \quad \{8, 17\}. \end{aligned}$$

Every element of $\mathbb{Z}_{18}$ has now been assigned to a coset, so these nine cosets form a partition of $\mathbb{Z}_{18}$.

(b) We take the subgroup H as the first coset:

$H = \{0, 6, 12\}$.

This does not contain the element 1, so we take $1 + H$ as the second coset:

$1 + H = \{1 +_{18} 0, 1 +_{18} 6, 1 +_{18} 12\} = \{1, 7, 13\}$.

These two cosets do not contain the element 2, so we take $2 + H$ as the third coset:

$2 + H = \{2 +_{18} 0, 2 +_{18} 6, 2 +_{18} 12\} = \{2, 8, 14\}$.

Continuing this process, using the elements 3, 4 and 5 in turn, we obtain the following cosets:

$\{3, 9, 15\}, \quad \{4, 10, 16\}, \quad \{5, 11, 17\}$.

Every element of $\mathbb{Z}_{18}$ has now been assigned to a coset, so these six cosets form a partition of $\mathbb{Z}_{18}$.

1.11 We take the subgroup H as the first coset:

$H = \{1, 12\}$.

This does not contain the element 2, so we take $2H$ as the second coset:

$2H = 2\{1, 12\} = \{2 \times_{13} 1, 2 \times_{13} 12\} = \{2, 11\}$.

These two cosets do not contain the element 3, so we take $3H$ as the third coset:

$3H = 3\{1, 12\} = \{3 \times_{13} 1, 3 \times_{13} 12\} = \{3, 10\}$.

Continuing this process, using the elements 4, 5 and 6 in turn, we obtain the following cosets:

$4H = \{4, 9\}, \quad 5H = \{5, 8\}, \quad 6H = \{6, 7\}$.

Every element of $\mathbb{Z}_{13}^*$ has now been assigned to a coset, so these six cosets form a partition of $\mathbb{Z}_{13}^*$.

2.1 It follows from Lagrange's Theorem (Theorem 2.1) that the possible orders of the subgroups are the positive divisors of n.

(a) The possible orders are 1, 2, 4, 5, 10 and 20.

(b) The possible orders are 1, 5 and 25.

(c) The possible orders are 1 and 29.

2.2 (a) In this case,

$n = $ order of group $S_4 = 24$,

$m = $ order of subgroup $A_4 = 12$.

There are two cosets of A_4 in S_4 (namely, A_4 itself and the set $(1\ 2)H$ of all odd permutations), so $k = 2$.

The equation $n = km$ is

$24 = 2 \times 12$.

(b) We saw in the solution to Example 1.5 that there are three cosets of $\{0, 3, 6\}$ in $\mathbb{Z}_9$.

In this case,

$n = $ order of the group $\mathbb{Z}_9 = 9$,

$m = $ order of the subgroup $\{0, 3, 6\} = 3$,

$k = $ number of cosets of $\{0, 3, 6\}$ in $\mathbb{Z}_9 = 3$.

So the equation $n = km$ is

$9 = 3 \times 3$.

2.3 In a finite group G,

index of H in $G = $ (order of G)/(order of H).

(a) $40/8 = 5$

(b) $45/3 = 15$

2.4 The following is a complete list of all the subgroups of A_4 together with their order.

Order	Subgroup of A_4
1	$\{e\}$
2	$\{e, (1\ 2)(3\ 4)\}$
2	$\{e, (1\ 3)(2\ 4)\}$
2	$\{e, (1\ 4)(2\ 3)\}$
3	$\langle(1\ 2\ 3)\rangle = \langle(1\ 3\ 2)\rangle = \{e, (1\ 2\ 3), (1\ 3\ 2)\}$
3	$\langle(1\ 2\ 4)\rangle = \langle(1\ 4\ 2)\rangle = \{e, (1\ 2\ 4), (1\ 4\ 2)\}$
3	$\langle(1\ 3\ 4)\rangle = \langle(1\ 4\ 3)\rangle = \{e, (1\ 3\ 4), (1\ 4\ 3)\}$
3	$\langle(2\ 3\ 4)\rangle = \langle(2\ 4\ 3)\rangle = \{e, (2\ 3\ 4), (2\ 4\ 3)\}$
4	$\{e, (1\ 2)(3\ 4), (1\ 3)(2\ 4), (1\ 4)(2\ 3)\}$
12	A_4

2.5 (a) Here g is a 4-cycle and so has order 4. The order of S_4 is $4! = 24$. Hence the order of g divides the order of S_4.

(b) Here g is a 3-cycle and so has order 3. The order of S_4 is $4! = 24$. Hence the order of g divides the order of S_4.

(c) The order of $\mathbb{Z}_9$ is 9.

The order of the element $g = 5$ is 9 because the identity element in $\mathbb{Z}_9$ is 0 and

$5 +_9 5 = 1, \quad 1 +_9 5 = 6,$

$6 +_9 5 = 2, \quad 2 +_9 5 = 7,$

$7 +_9 5 = 3, \quad 3 +_9 5 = 8,$

$8 +_9 5 = 4, \quad 4 +_9 5 = 0.$

(That is, we must add together nine 5s to obtain 0.)

Hence the order of g divides the order of $\mathbb{Z}_9$ because 9 divides 9.

(d) The order of $\mathbb{Z}_9$ is 9.

The order of the element $g = 6$ is 3 because the identity element in $\mathbb{Z}_9$ is 0 and

$6 +_9 6 = 3, \quad 3 +_9 6 = 0.$

Hence the order of g divides the order of $\mathbb{Z}_9$ because 3 divides 9.

2.6 (a) The group G has order 5, which is a prime, so G is cyclic.

(b) The identity element in the group is y because the row and column labelled y repeat the borders of the table. To verify that the other elements have order 5, we calculate their successive powers, using the information in the Cayley table:

$$v^2 = w, \qquad v^3 = w \circ v = z,$$
$$v^4 = z \circ v = x, \quad v^5 = x \circ v = y,$$

so v has order 5;

$$w^2 = x, \qquad w^3 = x \circ w = v,$$
$$w^4 = v \circ w = z, \quad w^5 = z \circ w = y,$$

so w has order 5;

$$x^2 = z, \qquad x^3 = z \circ x = w,$$
$$x^4 = w \circ x = v, \quad x^5 = v \circ x = y,$$

so x has order 5;

$$z^2 = v, \qquad z^3 = v \circ z = x,$$
$$z^4 = x \circ z = w, \quad z^5 = w \circ z = y,$$

so z has order 5.

(c) The group G has generators w, x, v and z.

The group $\mathbb{Z}_5$ has generators $1, 2, 3$ and 4.

Using the technique of matching powers of the generators w and 1, we obtain the following isomorphism.

$$\phi : G \longrightarrow \mathbb{Z}_5$$
$$w \longmapsto 1$$
$$w \circ w \longmapsto 1 +_5 1$$
$$w \circ w \circ w \longmapsto 1 +_5 1 +_5 1$$
$$w \circ w \circ w \circ w \longmapsto 1 +_5 1 +_5 1 +_5 1$$
$$w \circ w \circ w \circ w \circ w \longmapsto 1 +_5 1 +_5 1 +_5 1 +_5 1$$

This simplifies to the following isomorphism.

$$\phi : G \longrightarrow \mathbb{Z}_5$$
$$y \longmapsto 0$$
$$w \longmapsto 1$$
$$x \longmapsto 2$$
$$v \longmapsto 3$$
$$z \longmapsto 4$$

(There are three other isomorphisms, obtained by mapping $w \longmapsto 2$, $w \longmapsto 3$ and $w \longmapsto 4$.)

2.7 (a) Since $|G| = 14$, the possible orders of proper subgroups of G are 1, 2 and 7, by Lagrange's Theorem.

The trivial subgroup has order 1, which is certainly cyclic, and since 2 and 7 are primes, all the proper subgroups of G are cyclic, by Corollary 2 to Lagrange's Theorem.

(b) Since $|G| = 15$, the possible orders of proper subgroups of G are 1, 3 and 5, by Lagrange's Theorem.

The trivial subgroup has order 1, which is cyclic, and since 3 and 5 are primes, all the proper subgroups of G are cyclic, by Corollary 2 to Lagrange's Theorem.

(c) Since $|G| = pq$, where both p and q are primes, the possible orders of proper subgroups of G are 1, p and q, by Lagrange's Theorem.

The trivial subgroup has order 1, which is cyclic, and since p and q are primes, all the proper subgroups of G are cyclic, by Corollary 2 to Lagrange's Theorem.

Remark Parts (a) and (b) are special cases of part (c).

2.8 (a) It follows from Lagrange's Theorem that the possible orders of the subgroups of a group G are the positive divisors of $|G|$.

(i) The possible orders are 1 and 17.

(ii) The possible orders are 1, 2, 3, 4, 6, 9, 12, 18 and 36.

(iii) The possible orders are 1, 2, 4, 7, 8, 14, 28 and 56.

(iv) The possible orders are 1 and 59.

(b) Groups of prime order are necessarily cyclic by Corollary 2 to Lagrange's Theorem, so the groups of order 17 and 59 are cyclic.

(c) The group G might be cyclic in all cases. For each positive integer n there is a cyclic group of order n, namely $\mathbb{Z}_n$.

Remark There are also groups of order 36 and 56 which are non-cyclic.

(d) The group $\mathbb{Z}_{59}$ is a cyclic group of prime order. By Corollary 2 to Lagrange's Theorem, each element, except the identity, generates the whole group. So, in particular, 7 is a generator of $\mathbb{Z}_{59}$; that is, $\langle 7 \rangle = \mathbb{Z}_{59}$. Thus 7 generates a subgroup of order 59.

2.9 (a)

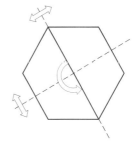

The following symmetries of the hexagon map the diagonal to itself.

The identity element,
rotation through π,
reflection in the diagonal,
reflection in the line perpendicular to the diagonal.

The subgroup has order 4.

(b)

order 1
(the identity)

order 2
(the identity and one reflection)

order 3
(cyclic—generated by
rotation through $2\pi/3$)

order 6
(the six symmetries of the triangle)

(You may have drawn different figures with the
required number of symmetries.)

3.1 (a) The orders of the elements are as follows.

Element	e	$(1\ 3)$	$(2\ 5)$	$(1\ 3)(2\ 5)$
Order	1	2	2	2

Since the group contains no element of order 4, it
must be isomorphic to K_4.

(b) The orders of the elements are as follows.

Element	e	$(2\ 3\ 4\ 6)$	$(2\ 4)(3\ 6)$	$(2\ 6\ 4\ 3)$
Order	1	4	2	4

Since the group contains an element or order 4, it
must be isomorphic to C_4.

Remark The group is

$$\langle(2\ 3\ 4\ 6)\rangle = \langle(2\ 6\ 4\ 3)\rangle.$$

3.2 There are many possible answers.

(a) Any cyclic subgroup of S_6 of order 6 will be
isomorphic to C_6. One possibility is the subgroup
generated by the element $(1\ 2\ 3\ 4\ 5\ 6)$. In full,

$$\langle(1\ 2\ 3\ 4\ 5\ 6)\rangle$$
$$= \{e, (1\ 2\ 3\ 4\ 5\ 6), (1\ 3\ 5)(2\ 4\ 6),$$
$$(1\ 4)(2\ 5)(3\ 6), (1\ 5\ 3)(2\ 6\ 4), (1\ 6\ 5\ 4\ 3\ 2)\}.$$

(b) Any non-Abelian subgroup of S_6 of order 6 is
isomorphic to $S(\triangle)$. One possibility is the subgroup

$$\{e, (2\ 3), (2\ 6), (3\ 6), (2\ 3\ 6), (2\ 6\ 3)\}$$

obtained by labelling the vertices of the triangle with
the symbols 2, 3 and 6.

3.3 The tables in parts (a), (b) and (c) are
symmetric about the leading diagonals, so the groups
in parts (a), (b) and (c) are Abelian, whereas the
group in part (d) is non-Abelian.

(a) This group has 7 elements of order 2 and so
belongs to class 2.

(b) This group has 3 elements of order 2 and so
belongs to class 3.

(c) This group has only 1 element of order 2 and so
belongs to class 1.

Remark The table shows that this group is cyclic
because there is a pattern of diagonal stripes.

(d) This group is non-Abelian and has only 1
element of order 2 and so belongs to class 5.

3.4 The table for P is symmetric about the leading
diagonal, but the table for Q is not, so P is Abelian
and Q is non-Abelian.

(a) The group P has 3 elements of order 2 and so
belongs to class 3.

(b) The subgroup $\{e, a, b, c\}$ of P contains an
element of order 4 (a or c) and so belongs to the
isomorphism class of C_4.

(c) The group Q has only 1 element of order 2 and
so belongs to class 5.

(d) The subgroup $\{e, i, w, x\}$ of Q contains an
element or order 4 (i or x) and so belongs to the
isomorphism class of C_4.

3.5 The subgroup $\{e, b, p, r\}$ of P is isomorphic
to K_4.

Remark You can see from the following group table
of these elements that it *is* a subgroup.

$\circ$	e	b	p	r
e	e	b	p	r
b	b	e	r	p
p	p	r	e	b
r	r	p	b	e

Since the subgroup is of order 4 but contains no
element of order 4, it is isomorphic to K_4.

3.6 Suppose that A_4 has a subgroup H of order 6.

As A_4 has no element of order 6, H is isomorphic to
the non-Abelian group $S(\triangle)$. Thus H must
contain e, two elements of order 3 and three elements
of order 2.

Now A_4 contains only three elements of order 2, so
these must all be contained in H. However, these
three elements of order 2, along with e, form a
subgroup of A_4, and also of H, of order 4.

This contradicts Lagrange's Theorem as 4 does not
divide 6. Thus A_4 cannot have a subgroup of order 6.

4.1 We use Strategy 4.1 for partitioning a group into right cosets.

(a) We take the subgroup H as the first right coset:
$$H = \{e, s\}.$$
This does not contain the element a, so we take Ha as the second right coset:
$$Ha = \{e \circ a, s \circ a\} = \{a, r\}.$$
These two right cosets do not contain the element b, so we take Hb as the third right coset:
$$Hb = \{e \circ b, s \circ b\} = \{b, u\}.$$
These three right cosets do not contain the element c, so we take Hc as the fourth right coset:
$$Hc = \{e \circ c, s \circ c\} = \{c, t\}.$$
Every element of $S(\square)$ has now been assigned to a right coset, so the partition is
$$\{e, s\} \quad \{a, r\}, \quad \{b, u\}, \quad \{c, t\}.$$

(b) We take the subgroup H as the first right coset:
$$H = \{e, b, r, t\}.$$
This does not contain the element a, so we take Ha as the second right coset:
$$Ha = \{e \circ a, b \circ a, r \circ a, t \circ a\}$$
$$= \{a, c, u, s\}.$$
Every element of $S(\square)$ has now been assigned to a right coset, so the partition is
$$\{e, b, r, t\}, \quad \{a, c, u, s\}.$$

4.2 We take H as the first right coset:
$$H = \{e, (1\ 2)\}.$$
This does not contain the element $(1\ 3)$, so we take $H(1\ 3)$ as the second right coset:
$$H(1\ 3) = \{e \circ (1\ 3), (1\ 2) \circ (1\ 3)\}$$
$$= \{(1\ 3), (1\ 3\ 2)\}.$$
These two cosets do not contain the element $(2\ 3)$, so we take $H(2\ 3)$ as the third right coset:
$$H(2\ 3) = \{e \circ (2\ 3), (1\ 2) \circ (2\ 3)\}$$
$$= \{(2\ 3), (1\ 2\ 3)\}.$$
Every element of S_3 has now been assigned to a right coset, so the partition is
$$\{e, (1\ 2)\}, \quad \{(1\ 3), (1\ 3\ 2)\}, \quad \{(2\ 3), (1\ 2\ 3)\}.$$

4.3 By definition,
$$Hg = \{h \circ g : h \in H\}.$$
If $g \in H$, then, for each $h \in H$,
$$h \circ g \in H \quad \text{(closure in } H\text{)},$$
so
$$Hg \subseteq H. \tag{S.1}$$
The subgroup H is finite, so it follows that $Hg = H$ because $Hg \subseteq H$, and the cosets H and Hg have the same number of elements, by property 1 of right cosets.

Conversely, if $Hg = H$, then each element of Hg belongs to H. In particular, since $e \in H$, the element $e \circ g = g$ belongs to Hg, so
$$g \in H.$$

4.4 (a) All the right cosets of H contain the same number of elements, by property 1 of right cosets.

It follows that there is just one other right coset—namely $\{r, s, t, u\}$.

So the partition into right cosets is
$$\{e, a, b, c\}, \quad \{r, s, t, u\}.$$
This is the same as the partition into left cosets of H, found in Exercise 1.4(a).

(b) All the right cosets of H contain the same number of elements, by property 1 of right cosets.

Thus each right coset of the subgroup $\{e\}$ has just one element.

So the partition into right cosets is
$$\{e\}, \{a\}, \{b\}, \{c\}, \{r\}, \{s\}, \{t\}, \{u\}.$$
This is the same as the partition into left cosets of H, found in Exercise 1.4(b).

(c) The subgroup $S(\square)$ is the whole group, so is the only right coset.

So the partition into right cosets is $S(\square)$.

This is the same as the partition into left cosets of H, found in Exercise 1.4(c).

4.5 (a) The left coset $(1\ 2)(3\ 4)H$ is
$$(1\ 2)(3\ 4)H = \{(1\ 2)(3\ 4) \circ e, (1\ 2)(3\ 4) \circ (1\ 2\ 3),$$
$$(1\ 2)(3\ 4) \circ (1\ 3\ 2)\}$$
$$= \{(1\ 2)(3\ 4), (2\ 4\ 3), (1\ 4\ 3)\}.$$
The right coset $H(1\ 2)(3\ 4)$ is
$$H(1\ 2)(3\ 4) = \{e \circ (1\ 2)(3\ 4), (1\ 2\ 3) \circ (1\ 2)(3\ 4),$$
$$(1\ 3\ 2) \circ (1\ 2)(3\ 4)\}$$
$$= \{(1\ 2)(3\ 4), (1\ 3\ 4), (2\ 3\ 4)\}.$$
The left and right cosets $(1\ 2)(3\ 4)H$ and $H(1\ 2)(3\ 4)$ are not the same; hence H is not a normal subgroup of A_4.

(b) The left cosets of N in A_4 are
$$N = \{e, (1\ 2)(3\ 4), (1\ 3)(2\ 4), (1\ 4)(2\ 3)\},$$
$$(1\ 2\ 3)N = \{(1\ 2\ 3), (1\ 3\ 4), (2\ 4\ 3), (1\ 4\ 2)\},$$
$$(1\ 3\ 2)N = \{(1\ 3\ 2), (2\ 3\ 4), (1\ 2\ 4), (1\ 4\ 3)\}.$$
The right cosets of N in A_4 are
$$N = \{e, (1\ 2)(3\ 4), (1\ 3)(2\ 4), (1\ 4)(2\ 3)\},$$
$$N(1\ 2\ 3) = \{(1\ 2\ 3), (2\ 4\ 3), (1\ 4\ 2), (1\ 3\ 4)\},$$
$$N(1\ 3\ 2) = \{(1\ 3\ 2), (1\ 4\ 3), (2\ 3\ 4), (1\ 2\ 4)\}.$$
The partitions of A_4 into left and right cosets of N are the same; hence N is a normal subgroup of A_4.

4.6 (a) For any element $g \in G$,
$$g\{e\} = \{g \circ e\} = \{g\}$$
and
$$\{e\}g = \{e \circ g\} = \{g\}.$$
The partitions of G into left and right cosets of $\{e\}$ are the same; hence $\{e\}$ is a normal subgroup of G.

(b) Recall that, for any element $g \in G$,
$$gG = G \quad \text{(Exercise 1.3)}$$
and
$$Gg = G \quad \text{(Exercise 4.3).}$$
It follows that
$$gG = Gg;$$
hence the group G is a normal subgroup of itself.

4.7 (a) The left cosets of H in G are
$$H = \{e, b, s, u\},$$
$$rH = \{r \circ e, r \circ b, r \circ s, r \circ u\}$$
$$= \{r, t, c, a\}.$$
The right cosets of H in G are
$$H = \{e, b, s, u\},$$
$$Hr = \{e \circ r, b \circ r, s \circ r, u \circ r\}$$
$$= \{r, t, a, c\}.$$
The partitions of G into left and right cosets of H are the same; hence H is a normal subgroup of G.

(b) The left cosets of H in G are
$$H = \{0, 2, 4, 6\},$$
$$1 + H = \{1 +_8 0, 1 +_8 2, 1 +_8 4, 1 +_8 6\}$$
$$= \{1, 3, 5, 7\}.$$
The right cosets of H in G are
$$H = \{0, 2, 4, 6\},$$
$$H + 1 = \{0 +_8 1, 2 +_8 1, 4 +_8 1, 6 +_8 1\}$$
$$= \{1, 3, 5, 7\}.$$
The partitions of G into left and right cosets of H are the same; hence H is a normal subgroup of G.

(c) The left cosets of A_3 in S_3 are
$$A_3 = \{e, (1\ 2\ 3), (1\ 3\ 2)\},$$
$$(1\ 2)A_3 = \{(1\ 2) \circ e, (1\ 2) \circ (1\ 2\ 3), (1\ 2) \circ (1\ 3\ 2)\}$$
$$= \{(1\ 2), (2\ 3), (1\ 3)\}.$$
The right cosets of A_3 in S_3 are
$$A_3 = \{e, (1\ 2\ 3), (1\ 3\ 2)\},$$
$$A_3(1\ 2) = \{e \circ (1\ 2), (1\ 2\ 3) \circ (1\ 2), (1\ 3\ 2) \circ (1\ 2)\}$$
$$= \{(1\ 2), (1\ 3), (2\ 3)\}.$$
The partitions of S_3 into left and right cosets of A_3 are the same; hence A_3 is a normal subgroup of S_3.

4.8 We use Strategy 4.1.

(a) We take the subgroup H as the first (right) coset:
$$H = \{e, w\}.$$
This does not contain the element i, so we take Hi as the second coset:
$$Hi = \{e \circ i, w \circ i\} = \{i, x\}.$$
These two cosets do not contain the element j, so we take Hj as the third coset:
$$Hj = \{e \circ j, w \circ j\} = \{j, y\}.$$
These three cosets do not contain the element k, so we take Hk as the fourth coset:
$$Hk = \{e \circ k, w \circ k\} = \{k, z\}.$$
Every element of Q has now been assigned to a coset, so the right partition of Q into cosets of H is
$$\{e, w\}, \quad \{i, x\}, \quad \{j, y\}, \quad \{k, z\}.$$
This is the same as the left partition that we found in Exercise 1.8(a). It follows that $\{e, w\}$ is a normal subgroup of $(Q, \circ)$.

(b) We take the subgroup H as the first (right) coset:
$$H = \{e, i, w, x\}.$$
This does not contain the element j, so we take Hj as the second coset:
$$Hj = \{e \circ j, i \circ j, w \circ j, x \circ j\}$$
$$= \{j, k, y, z\}.$$
Every element of Q has now been assigned to a coset, so the right partition of Q into cosets of H is
$$\{e, i, w, x\}, \quad \{j, k, y, z\}.$$
This is the same as the left partition that we found in Exercise 1.8(b). It follows that $\{e, i, w, x\}$ is a normal subgroup of $(Q, \circ)$.

4.9 (a) The coset $H(1\ 2)$ consists of all the elements of S_4 of the form
$$h \circ (1\ 2), \quad \text{where } h \in H.$$
Let h be any element of H so that $h(1) = 1$, and consider the effect of the composite $h \circ (1\ 2)$ on the symbol 2:
$$2 \xrightarrow{(1\ 2)} 1 \xrightarrow{h} 1.$$
So each element of the coset $H(1\ 2)$ maps 2 to 1. Now the coset $H(1\ 2)$ contains six elements (because H has order 6) and there are just six elements of S_4 which map 2 to 1, namely
$$(2\ 1), \quad (2\ 1)(3\ 4), \quad (2\ 1\ 3),$$
$$(2\ 1\ 4), \quad (2\ 1\ 3\ 4), \quad (2\ 1\ 4\ 3).$$
So these are the six elements of $H(1\ 2)$.

(b) Similarly, the coset $H(1\ 3)$ consists of the six elements of S_4 which map 3 to 1, namely

$$(3\ 1),\quad (3\ 1)(2\ 4),\quad (3\ 1\ 2),$$
$$(3\ 1\ 4),\quad (3\ 1\ 2\ 4),\quad (3\ 1\ 4\ 2),$$

and $H(1\ 4)$ consists of the six elements of S_4 which map 4 to 1, namely

$$(4\ 1),\quad (4\ 1)(2\ 3),\quad (4\ 1\ 2),$$
$$(4\ 1\ 3),\quad (4\ 1\ 2\ 3),\quad (4\ 1\ 3\ 2).$$

The four right cosets H, $H(1\ 2)$, $H(1\ 3)$ and $H(1\ 4)$ contain all $4! = 24$ elements of S_4 and so complete the right partition of S_4 into cosets of H.

(c) The left and right partitions of S_4 into cosets of H are not the same; for example, the right coset $H(1\ 2)$ found above is not the same as the left coset $(1\ 2)H$ found in Exercise 1.9(a), so H is not a normal subgroup of S_4.

4.10 (a) The group $\mathbb{Z}_{16}$ is Abelian. By Theorem 4.2, all its subgroups are normal. In particular, $H = \{0, 4, 8, 12\}$ is a normal subgroup of $\mathbb{Z}_{16}$.

(b) The group A_5 is a subgroup of index 2 in S_5. By Theorem 4.3, all subgroups of index 2 are normal. In particular, A_5 is a normal subgroup of S_5. (Alternatively, we could quote the corollary to Theorem 4.3, with $n = 5$.)

(c) The group $H = \{e, k, w, z\}$ is a subgroup of index 2 in $(Q, \circ)$. By Theorem 4.3, all subgroups of index 2 are normal. In particular, H is a normal subgroup of $(Q, \circ)$.

5.1 (a)
$$\begin{aligned}
rH \cdot H &= \{r, t\} \cdot \{e, b\} \\
&= \{r \circ e, r \circ b, t \circ e, t \circ b\} \\
&= \{r, t, t, r\} \\
&= \{r, t\} = rH
\end{aligned}$$

(b)
$$\begin{aligned}
sH \cdot rH &= \{s, u\} \cdot \{r, t\} \\
&= \{s \circ r, s \circ t, u \circ r, u \circ t\} \\
&= \{a, c, c, a\} \\
&= \{a, c\} = aH
\end{aligned}$$

(c)
$$\begin{aligned}
aH \cdot aH &= \{a, c\} \cdot \{a, c\} \\
&= \{a \circ a, a \circ c, c \circ a, c \circ c\} \\
&= \{b, e, e, b\} \\
&= \{e, b\} = H
\end{aligned}$$

Remark In this example, each of the composites is a coset of H.

5.2 (a)
$$\begin{aligned}
aH \cdot aH &= \{a, s\} \cdot \{a, s\} \\
&= \{a \circ a, a \circ s, s \circ a, s \circ s\} \\
&= \{b, t, r, e\}
\end{aligned}$$

(b)
$$\begin{aligned}
bH \cdot cH &= \{b, t\} \cdot \{c, u\} \\
&= \{b \circ c, b \circ u, t \circ c, t \circ u\} \\
&= \{a, s, u, c\}
\end{aligned}$$

(c)
$$\begin{aligned}
cH \cdot bH &= \{c, u\} \cdot \{b, t\} \\
&= \{c \circ b, c \circ t, u \circ b, u \circ t\} \\
&= \{a, s, s, a\} = \{a, s\} = aH
\end{aligned}$$

Remark In parts (a) and (b), the composites contain four elements, so they are not cosets of H.

5.3 (a)
$$\begin{aligned}
H + H &= \{0, 3\} + \{0, 3\} \\
&= \{0 +_6 0, 0 +_6 3, 3 +_6 0, 3 +_6 3\} \\
&= \{0, 3, 3, 0\} \\
&= \{0, 3\} = H
\end{aligned}$$

(b)
$$\begin{aligned}
(2 + H) + (1 + H) &= \{2, 5\} + \{1, 4\} \\
&= \{2 +_6 1, 2 +_6 4, 5 +_6 1, 5 +_6 4\} \\
&= \{3, 0, 0, 3\} \\
&= \{0, 3\} = H
\end{aligned}$$

(c)
$$\begin{aligned}
(2 + H) + (2 + H) &= \{2, 5\} + \{2, 5\} \\
&= \{2 +_6 2, 2 +_6 5, 5 +_6 2, 5 +_6 5\} \\
&= \{4, 1, 1, 4\} \\
&= \{1, 4\} = 1 + H
\end{aligned}$$

Remark In this example, each of the composites is a coset of H.

5.4 (a) We saw in Exercise 1.1(a) that the left cosets of $H = \{e, s\}$ in $S(\square)$ are

$$H = \{e, s\}, \quad aH = \{a, t\},$$
$$bH = \{b, u\}, \quad cH = \{c, r\}.$$

We now write out the group table for $S(\square)$, listing the border elements in the order e, s, a, t, b, u, c, r.

		H		aH		bH		cH	
$\circ$		e	s	a	t	b	u	c	r
H	e	e	s	a	t	b	u	c	r
	s	s	e	r	c	u	b	t	a
aH	a	a	t	b	u	c	r	e	s
	t	t	a	s	e	r	c	u	b
bH	b	b	u	c	r	e	s	a	t
	u	u	b	t	a	s	e	r	c
cH	c	c	r	e	s	a	t	b	u
	r	r	c	u	b	t	a	s	e

This table does not block; for example,

$$\begin{aligned}
H \cdot aH &= \{e, s\} \cdot \{a, t\} \\
&= \{e \circ a, e \circ t, s \circ a, s \circ t\} \\
&= \{a, t, r, c\},
\end{aligned}$$

which is not a coset.

Remark It is not necessary to write out the whole table; it is sufficient to find two cosets whose set composite is not a coset.

(b) We saw in Exercise 1.1(b) that the cosets of $H = \{e, b, r, t\}$ in $S(\square)$ are

$$H = \{e, b, r, t\}, \quad aH = \{a, c, s, u\}.$$

We now write out the group table for $S(\square)$, listing the border elements in the order e, b, r, t, a, c, s, u.

		H				aH		
$\circ$	e	b	r	t	a	c	s	u
e	e	b	r	t	a	c	s	u
b	b	e	t	r	c	a	u	s
r	r	t	e	b	u	s	c	a
t	t	r	b	e	s	u	a	c
a	a	c	s	u	b	e	t	r
c	c	a	u	s	e	b	r	t
s	s	u	a	c	r	t	e	b
u	u	s	c	a	t	r	b	e

The row-border is labelled H for e, b, r, t and aH for a, c, s, u.

The table blocks as follows.

$\cdot$	H	aH
H	H	aH
aH	aH	H

5.5 (a) The group $\mathbb{Z}_8$ is Abelian, so all its subgroups are normal, by Theorem 4.2.

(b) By the result in Frame 14, the order of the quotient group $\mathbb{Z}_8/\{0,4\}$ is

(order of $\mathbb{Z}_8$)/(order of $\{0,4\}$) $= 8/2 = 4$.

The four cosets of H in $\mathbb{Z}_8$ are

$0 + H = \{0, 4\} = 4 + H$,
$1 + H = \{1, 5\} = 5 + H$,
$2 + H = \{2, 6\} = 6 + H$,
$3 + H = \{3, 7\} = 7 + H$.

By Theorem 5.1 (Frame 12), the rule for composing cosets is

$(a + H) + (b + H) = (a +_8 b) + H$.

For example,

$H + (1 + H) = (0 + H) + (1 + H)$
$\qquad\qquad = (0 +_8 1) + H$
$\qquad\qquad = 1 + H$,

$(2 + H) + (3 + H) = (2 +_8 3) + H$
$\qquad\qquad = 5 + H = 1 + H$.

Continuing in the same way, we obtain the following group table for the quotient group $\mathbb{Z}_8/H$.

$+$	H	$1 + H$	$2 + H$	$3 + H$
H	H	$1 + H$	$2 + H$	$3 + H$
$1 + H$	$1 + H$	$2 + H$	$3 + H$	H
$2 + H$	$2 + H$	$3 + H$	H	$1 + H$
$3 + H$	$3 + H$	H	$1 + H$	$2 + H$

The identity element is H.

The table of inverses is as follows.

Element	H	$1 + H$	$2 + H$	$3 + H$
Inverse	H	$3 + H$	$2 + H$	$1 + H$

(c) A group of order 4 is isomorphic either to $\mathbb{Z}_4$ or to K_4. However, in $\mathbb{Z}_8/H$ not all the elements are self-inverse, so $\mathbb{Z}_8/H \not\cong K_4$. Hence

$\mathbb{Z}_8/H \cong \mathbb{Z}_4$.

Alternatively,

$(1 + H) + (1 + H) = 2 + H$,
$(2 + H) + (1 + H) = 3 + H$,
$(3 + H) + (1 + H) = 4 + H = H$.

Hence $\mathbb{Z}_8/H$ is a cyclic group of order 4, generated by $1 + H$; that is,

$\mathbb{Z}_8/H \cong \mathbb{Z}_4$.

5.6 (a) The group $\mathbb{Z}_{12}$ is Abelian, so all its subgroups are normal, by Theorem 4.2.

(b) By the result in Frame 14, the order of the quotient group $\mathbb{Z}_{12}/\{0,6\}$ is

(order of $\mathbb{Z}_{12}$)/(order of $\{0,6\}$) $= 12/2 = 6$.

The six cosets of H in $\mathbb{Z}_{12}$ are

$0 + H = \{0, 6\}\ \ = 6 + H$,
$1 + H = \{1, 7\}\ \ = 7 + H$,
$2 + H = \{2, 8\}\ \ = 8 + H$,
$3 + H = \{3, 9\}\ \ = 9 + H$,
$4 + H = \{4, 10\} = 10 + H$,
$5 + H = \{5, 11\} = 11 + H$.

By Theorem 5.1 (Frame 12), the rule for composing cosets is

$(a + H) + (b + H) = (a +_{12} b) = H$.

For example,

$H + (1 + H) = (0 + H) + (1 + H)$
$\qquad\qquad = (0 +_{12} 1) + H$
$\qquad\qquad = 1 + H$,

$(2 + H) + (4 + H) = (2 +_{12} 4) + H$
$\qquad\qquad = 6 + H = H$,

$(3 + H) + (5 + H) = (3 +_{12} 5) + H$
$\qquad\qquad = 8 + H = 2 + H$.

Continuing in the same way, we obtain the following group table for the quotient group $\mathbb{Z}_{12}/\{0,6\}$.

$+$	H	$1 + H$	$2 + H$	$3 + H$	$4 + H$	$5 + H$
H	H	$1 + H$	$2 + H$	$3 + H$	$4 + H$	$5 + H$
$1 + H$	$1 + H$	$2 + H$	$3 + H$	$4 + H$	$5 + H$	H
$2 + H$	$2 + H$	$3 + H$	$4 + H$	$5 + H$	H	$1 + H$
$3 + H$	$3 + H$	$4 + H$	$5 + H$	H	$1 + H$	$2 + H$
$4 + H$	$4 + H$	$5 + H$	H	$1 + H$	$2 + H$	$3 + H$
$5 + H$	$5 + H$	H	$1 + H$	$2 + H$	$3 + H$	$4 + H$

(c) A group of order 6 is isomorphic either to $\mathbb{Z}_6$ or to $S(\triangle)$. However, the group $\mathbb{Z}_{12}/H$ is Abelian, so it is not isomorphic to $S(\triangle)$. Hence

$\mathbb{Z}_{12}/H \cong \mathbb{Z}_6$.

Alternatively,

$$(1 + H) + (1 + H) = 2 + H,$$
$$(2 + H) + (1 + H) = 3 + H,$$
$$(3 + H) + (1 + H) = 4 + H,$$
$$(4 + H) + (1 + H) = 5 + H,$$
$$(5 + H) + (1 + H) = 6 + H = H.$$

Hence $\mathbb{Z}_{12}/H$ is a cyclic group of order 6, generated by $1 + H$; that is

$$\mathbb{Z}_{12}/H \cong \mathbb{Z}_6.$$

5.7 First we consider

$$N = \{e, a, b, c, d, f, g, h, i, j, k, l\}.$$

(a) The cosets of N in G are

$$N = \{e, a, b, c, d, f, g, h, i, j, k, l\},$$
$$zN = \{z, y, x, w, u, v, s, t, q, r, o, p\}.$$

(b) The group table for G/N is as follows.

	N	zN
N	N	zN
zN	zN	N

(c) The elements N and zN are both self-inverse.

(d) The quotient group G/N has order 2 and the element zN has order 2, so G/N is cyclic and therefore isomorphic to $\mathbb{Z}_2$.

Next we consider $N = \{e, a, b, c\}$.

(a) The cosets of N in G are

$$N = \{e, a, b, c\}, \quad dN = \{d, f, g, h\},$$
$$iN = \{i, j, k, l\}, \quad zN = \{z, y, x, w\},$$
$$vN = \{v, u, t, s\}, \quad rN = \{r, q, p, o\}.$$

(b) The group table for G/N is as follows.

	N	dN	iN	zN	vN	rN
N	N	dN	iN	zN	vN	rN
dN	dN	iN	N	rN	zN	vN
iN	iN	N	dN	vN	rN	zN
zN	zN	vN	rN	N	dN	iN
vN	vN	rN	zN	iN	N	dN
rN	rN	zN	vN	dN	iN	N

(The blocking can be checked from the table on page 14.)

(c) From the above table, we see that

N, zN, vN and rN are self-inverse,
iN and dN are inverses of each other.

(d) A group of order 6 is isomorphic either to $\mathbb{Z}_6$ or to $S(\triangle)$.

The table for G/N is not symmetric about the main diagonal, so the quotient group is non-Abelian and so it is non-cyclic.

Hence

$$G/N \cong S(\triangle).$$

5.8 (a) $\{e, w\} \cdot \{i, x\}$
$$= \{e \circ i, e \circ x, w \circ i, w \circ x\}$$
$$= \{i, x, x, i\}$$
$$= \{i, x\}$$

(b) $\{j, y\} \cdot \{j, y\}$
$$= \{j \circ j, j \circ y, y \circ j, y \circ y\}$$
$$= \{w, e, e, w\}$$
$$= \{e, w\}$$

(c) $\{k, z\} \cdot \{i, x\}$
$$= \{k \circ i, k \circ x, z \circ i, z \circ x\}$$
$$= \{j, y, y, j\}$$
$$= \{j, y\}$$

5.9 (a) $iH \cdot jH = \{i, x\} \cdot \{j, y\}$
$$= \{i \circ j, i \circ y, x \circ j, x \circ y\}$$
$$= \{k, z, z, k\}$$
$$= \{k, z\}$$
$$= kH$$

(b) $jH \cdot jH = \{j, y\} \cdot \{j, y\}$
$$= \{j \circ j, j \circ y, y \circ j, y \circ y\}$$
$$= \{w, e, e, w\}$$
$$= \{e, w\}$$
$$= H$$

(c) $jH \cdot kH = \{j, y\} \cdot \{k, z\}$
$$= \{j \circ k, j \circ z, y \circ k, y \circ z\}$$
$$= \{i, x, x, i\}$$
$$= \{i, x\}$$

Hence

$$iH \cdot (jH \cdot kH) = \{i, x\} \cdot \{i, x\}$$
$$= \{i \circ i, i \circ x, x \circ i, x \circ x\}$$
$$= \{w, e, e, w\}$$
$$= \{e, w\}$$
$$= H.$$

5.10 We saw in Exercise 1.8(a) that the cosets of $H = \{e, w\}$ in $(Q, \circ)$ are

$$H = \{e, w\}, \quad iH = \{i, x\},$$
$$jH = \{j, y\}, \quad kH = \{k, z\}.$$

(a) We use Theorem 5.1 to calculate the entries in the group table of the quotient group; for example,

$$iH \cdot jH = (i \circ j)H = kH,$$
$$kH \cdot kH = (k \circ k)H = wH = H \quad \text{(since } w \in H\text{)},$$

and so on.

We thus obtain the following group table.

·	H	iH	jH	kH
H	H	iH	jH	kH
iH	iH	H	kH	jH
jH	jH	kH	H	iH
kH	kH	jH	iH	H

(b) Since Q/H is non-cyclic, it belongs to the isomorphism class of non-cyclic groups of order 4: it is isomorphic to $S(\square)$.

5.11 (a) The group $\mathbb{Z}_{13}^*$ is Abelian.

It follows from Theorem 4.2 that every subgroup of $\mathbb{Z}_{13}^*$ is normal; in particular, H is normal.

(b) By Theorem 5.1 (Frame 12), the rule for composing cosets is

$$aH \cdot bH = (a \times_{13} b)H.$$

For example,

$$4H \cdot 5H = (4 \times_{13} 5)H = 7H = 6H \quad \text{(since } 7 \in 6H\text{)},$$

and so on. We thus obtain the following group table.

·	H	$2H$	$3H$	$4H$	$5H$	$6H$
H	H	$2H$	$3H$	$4H$	$5H$	$6H$
$2H$	$2H$	$4H$	$6H$	$5H$	$3H$	H
$3H$	$3H$	$6H$	$4H$	H	$2H$	$5H$
$4H$	$4H$	$5H$	H	$3H$	$6H$	$2H$
$5H$	$5H$	$3H$	$2H$	$6H$	H	$4H$
$6H$	$6H$	H	$5H$	$2H$	$4H$	$3H$

(c) The quotient group is a cyclic group of order 6, generated by either $2H$ or $6H$.

For example,

$$(2H)^2 = 2H \cdot 2H = 4H,$$
$$(2H)^3 = 4H \cdot 2H = 5H,$$
$$(2H)^4 = 5H \cdot 2H = 3H,$$
$$(2H)^5 = 3H \cdot 2H = 6H,$$
$$(2H)^6 = 6H \cdot 2H = H.$$

Earlier we saw that any cyclic group of order n is isomorphic to $\mathbb{Z}_n$ (Unit GTA2, Theorem 3.1).

Thus the quotient group $\mathbb{Z}_{13}^*/\{1, 12\}$ is isomorphic to $\mathbb{Z}_6$.

6.1 There are six cosets of $6\mathbb{Z}$ in $\mathbb{Z}$, corresponding to the six remainders on division by 6, namely

$$0 + 6\mathbb{Z}, \quad 1 + 6\mathbb{Z}, \quad 2 + 6\mathbb{Z},$$
$$3 + 6\mathbb{Z}, \quad 4 + 6\mathbb{Z}, \quad 5 + 6\mathbb{Z}.$$

In constructing the group table, for brevity we omit '$+ 6\mathbb{Z}$' in each entry, writing the coset $k + 6\mathbb{Z}$ as just k, for $k = 0, 1, \ldots, 5$ as follows.

	0	1	2	3	4	5
0	0	1	2	3	4	5
1	1	2	3	4	5	0
2	2	3	4	5	0	1
3	3	4	5	0	1	2
4	4	5	0	1	2	3
5	5	0	1	2	3	4

With this abbreviated notation, we have the familiar group table for $\mathbb{Z}_6$ under addition modulo 6. We conclude that $\mathbb{Z}/6\mathbb{Z}$ is isomorphic to the cyclic group $\mathbb{Z}_6$.

To confirm that this group is cyclic, we look at the multiples of $1 + 6\mathbb{Z}$:

$$1 + 6\mathbb{Z},$$
$$2(1 + 6\mathbb{Z}) = (1 + 6\mathbb{Z}) + (1 + 6\mathbb{Z}) = 2 + 6\mathbb{Z},$$
$$3(1 + 6\mathbb{Z}) = 2(1 + 6\mathbb{Z}) + (1 + 6\mathbb{Z}) = 3 + 6\mathbb{Z},$$
$$4(1 + 6\mathbb{Z}) = 3(1 + 6\mathbb{Z}) + (1 + 6\mathbb{Z}) = 4 + 6\mathbb{Z},$$
$$5(1 + 6\mathbb{Z}) = 4(1 + 6\mathbb{Z}) + (1 + 6\mathbb{Z}) = 5 + 6\mathbb{Z},$$
$$6(1 + 6\mathbb{Z}) = 5(1 + 6\mathbb{Z}) + (1 + 6\mathbb{Z}) = 0 + 6\mathbb{Z}.$$

Thus $1 + 6\mathbb{Z}$ is an element of order 6 in a group of order 6 and is therefore a generator of the group, so $\mathbb{Z}/6\mathbb{Z}$ is cyclic.

6.2 (a) $\mathbb{Z}/5\mathbb{Z}$ is generated by each of its non-identity elements, namely $1 + 5\mathbb{Z}$, $2 + 5\mathbb{Z}$, $3 + 5\mathbb{Z}$ and $4 + 5\mathbb{Z}$.

(b) $\mathbb{Z}/6\mathbb{Z}$ is generated by $1 + 6\mathbb{Z}$ or $5 + 6\mathbb{Z}$.

$1 + n\mathbb{Z}$ seems a good candidate for a generator of $\mathbb{Z}/n\mathbb{Z}$, since 1 is a generator of $\mathbb{Z}_n$.

6.3 (a) There are only two different cosets in this list because

$$3.7 + \mathbb{Z}$$
$$= -1.3 + \mathbb{Z}$$
$$= \{\ldots, -2.3, -1.3, -0.3, 0.7, 1.7, 2.7, 3.7, \ldots\}$$

and

$$0.1 + \mathbb{Z}$$
$$= 39.1 + \mathbb{Z}$$
$$= -4.9 + \mathbb{Z}$$
$$= \{\ldots, -2.9, -1.9, -0.9, 0.1, 1.1, 2.1, 3.1, \ldots\}.$$

(b) The smallest positive member of the first coset is 0.7 and the smallest positive member of the secon coset is 0.1.

6.4 The fractional part of 3.1 is 0.1, so

$$3.1 + \mathbb{Z} = 0.1 + \mathbb{Z}.$$

The fractional part of -3.1 is 0.9, so

$$-3.1 + \mathbb{Z} = 0.9 + \mathbb{Z}.$$

The fractional part of $\sqrt{2}$ $(= 1.414\ldots)$ is $\sqrt{2} - 1$, so

$$\sqrt{2} + \mathbb{Z} = (\sqrt{2} - 1) + \mathbb{Z}.$$

6.5 **(a)** Since $0.3 + 0.42 = 0.72 \equiv 0.72 \pmod 1$,
$$(0.3 + \mathbb{Z}) + (0.42 + \mathbb{Z}) = 0.72 + \mathbb{Z}.$$

(b) Since $0.63 + 0.63 = 1.26 \equiv 0.26 \pmod 1$,
$$(0.63 + \mathbb{Z}) + (0.63 + \mathbb{Z}) = 0.26 + \mathbb{Z}.$$

(c) Since $0.63 + 0.37 = 1 \equiv 0 \pmod 1$,
$$(0.63 + \mathbb{Z}) + (0.37 + \mathbb{Z}) = 0 + \mathbb{Z} = \mathbb{Z}.$$

6.6 We check that the four group axioms hold.

G1 By the definition of $+_1$, $([0,1), +_1)$ is closed under $+_1$.

G2 The element $0 \in [0,1)$ is an identity element, since $x +_1 0 = 0 +_1 x = x$ for all $x \in [0,1)$.

G3 The element 0 of $[0,1)$ satisfies
$$0 +_1 0 = 0,$$
so it has inverse 0. If x is any other element of $[0,1)$, then $1 - x \in [0,1)$ and
$$x +_1 (1 - x) = (1 - x) +_1 x = 0,$$
so $1 - x$ is the inverse of x.

G4 Let x, y and z be elements of $[0,1)$.

Since we can obtain $x +_1 y$ by adding 0 or -1 to $x + y$, we can write
$$(x +_1 y) +_1 z$$
$$= (x + y + k) +_1 z, \quad k \in \{0, -1\}$$
$$= (x + y + k) + z + l, \quad l \in \{0, -1\}$$
$$= x + y + z + (l + k),$$
where we have used the fact that $+$ is commutative and associative. We can therefore write
$$(x +_1 y) +_1 z$$
$$= x + y + z + (\text{an integer}). \qquad (\text{S.2})$$
A similar calculation shows that
$$x +_1 (y +_1 z)$$
$$= x + y + z + (\text{an integer}). \qquad (\text{S.3})$$
Subtracting equation (S.3) from equation (S.2) and rearranging, we obtain
$$(x +_1 y) +_1 z$$
$$= x +_1 (y +_1 z) + (\text{an integer}). \qquad (\text{S.4})$$
Since $(x +_1 y) +_1 z$ and $x +_1 (y +_1 z)$ both lie in $[0,1)$, the integer in equation (S.4) must be 0, so we conclude that $+_1$ is associative.

Hence $([0,1), +_1)$ satisfies the four group axioms and so is a group.

6.7 **(a)** We have
$$3\mathbb{Z} = \{\ldots, -12, -9, -6, -3, 0, 3, 6, \ldots\},$$
so the coset $4.3 + 3\mathbb{Z}$ is
$$\{\ldots, -7.7, -4.7, -1.7, 1.3, 4.3, 7.3, \ldots\},$$

the coset $0.75 + 3\mathbb{Z}$ is
$$\{\ldots, -8.25, -5.25, -2.25, 0.75, 3.75, 6.75, \ldots\}$$
and the coset $-12.9 + 3\mathbb{Z}$ is
$$\{\ldots, -6.9, -3.9, -0.9, 2.1, 5.1, 8.1, \ldots\}.$$

(b) Each coset consists of real numbers that are evenly spaced 3 units apart. Precisely one of these real numbers therefore lies in the interval $[0,3)$. If we denote this number by r, then we can write the coset as $r + 3\mathbb{Z}$. It follows that the set of cosets of $3\mathbb{Z}$ in $\mathbb{Z}$ is
$$\{r + 3\mathbb{Z} : r \in [0,3)\}.$$

(c) Since $1 + 2.3 = 3.3$ and $3.3 + 3\mathbb{Z} = 0.3 + 3\mathbb{Z}$, we have
$$(1 + 3\mathbb{Z}) + (2.3 + 3\mathbb{Z}) = 0.3 + 3\mathbb{Z}.$$
Since $2.5 + 0 = 2.5$,
$$(2.5 + 3\mathbb{Z}) + (0 + 3\mathbb{Z}) = 2.5 + 3\mathbb{Z}.$$
Since $1.93 + 0.75 = 1.93 + 0.75 - 2.68$,
$$(1.93 + 3\mathbb{Z}) + (0.75 + 3\mathbb{Z}) = 2.68 + 3\mathbb{Z}.$$

(d) The composite of the two cosets $x + 3\mathbb{Z}$ and $y + 3\mathbb{Z}$ is obtained by adding the two numbers x and y, and then taking the remainder on division by 3; that is,
$$(x + 3\mathbb{Z}) + (y + 3\mathbb{Z}) = (x +_3 y) + 3\mathbb{Z},$$
where $+_3$ is addition of real numbers modulo 3.

(e) We conclude that $\mathbb{R}/3\mathbb{Z} \cong ([0,3), +_3)$.

6.8 The coset $(x, y)A$ is the set obtained by composing the element (x, y) with each element of the subgroup A; that is,
$$(x, y)A = \{(x, y) * (1, b) : b \in \mathbb{R}\}.$$

(a) $(2, 1)A = \{(2, 1) * (1, b) : b \in \mathbb{R}\}$
$$= \{(2, 2b + 1) : b \in \mathbb{R}\}$$
$$= \{(2, r) : r \in \mathbb{R}\}$$
(Since b ranges through all real values, $2b + 1$ takes all real values.)

(b) $(2, 3)A = \{(2, 3) * (1, b) : b \in \mathbb{R}\}$
$$= \{(2, 2b + 3) : b \in \mathbb{R}\}$$
$$= \{(2, r) : r \in \mathbb{R}\}$$
(Notice that $(2, 3)A = (2, 1)A$.)

(c) $(-2, \pi)A = \{(-2, \pi) * (1, b) : b \in \mathbb{R}\}$
$$= \{(-2, -2b + \pi) : b \in \mathbb{R}\}$$
$$= \{(-2, r) : r \in \mathbb{R}\}$$
(As b ranges through all real values, $-2b + \pi$ takes all real values.)

6.9 The element $(x, y) \in X$ belongs to the coset $(x, 0)A$.

(a) $(17, -3) \in (17, 0)A$

(b) $(-\sqrt{2}, \sqrt{2}) \in (-\sqrt{2}, 0)A$

(c) $(1, \tfrac{3}{4}) \in (1, 0)A = A$

6.10 (a) Using the formula
$$(x,0)A \cdot (y,0)A = (xy,0)A,$$
we obtain
$$(3.2,0)A \cdot (-3,0)A = (-9.6,0)A,$$
$$(\tfrac{7}{3},0)A \cdot (-\tfrac{2}{5},0)A = (-\tfrac{14}{15},0)A.$$

(b) We have
$$(3,-2) * (-2,3) = (-6,7)$$
and
$$(-6,7) \in (-6,0)A,$$
so
$$(3,-2) * (-2,3) \in (-6,0)A.$$

6.11 There are four cosets of $4\mathbb{Z}$ in $\mathbb{Z}$, namely
$$0 + 4\mathbb{Z} = \{\ldots, -8, -4, 0, 4, 8, \ldots\},$$
$$1 + 4\mathbb{Z} = \{\ldots, -7, -3, 1, 5, 9, \ldots\},$$
$$2 + 4\mathbb{Z} = \{\ldots, -6, -2, 2, 6, 10, \ldots\},$$
$$3 + 4\mathbb{Z} = \{\ldots, -5, -1, 3, 7, 11, \ldots\}.$$
These four cosets are the elements of the quotient group $\mathbb{Z}/4\mathbb{Z}$. The binary operation on this set is given by
$$(x + 4\mathbb{Z}) + (y + 4\mathbb{Z}) = (x +_4 y) + 4\mathbb{Z}.$$
The group table is therefore as follows.

+	$0+4\mathbb{Z}$	$1+4\mathbb{Z}$	$2+4\mathbb{Z}$	$3+4\mathbb{Z}$
$0+4\mathbb{Z}$	$0+4\mathbb{Z}$	$1+4\mathbb{Z}$	$2+4\mathbb{Z}$	$3+4\mathbb{Z}$
$1+4\mathbb{Z}$	$1+4\mathbb{Z}$	$2+4\mathbb{Z}$	$3+4\mathbb{Z}$	$0+4\mathbb{Z}$
$2+4\mathbb{Z}$	$2+4\mathbb{Z}$	$3+4\mathbb{Z}$	$0+4\mathbb{Z}$	$1+4\mathbb{Z}$
$3+4\mathbb{Z}$	$3+4\mathbb{Z}$	$0+4\mathbb{Z}$	$1+4\mathbb{Z}$	$2+4\mathbb{Z}$

Dropping '$+ 4\mathbb{Z}$' throughout, we can see that this group is isomorphic to the cyclic group $\mathbb{Z}_4$.

To confirm that this group is cyclic, look at the 'multiples' of $1 + 4\mathbb{Z}$:
$$1 + 4\mathbb{Z},$$
$$2(1 + 4\mathbb{Z}) = (1 + 4\mathbb{Z}) + (1 + 4\mathbb{Z}) = 2 + 4\mathbb{Z},$$
$$3(1 + 4\mathbb{Z}) = 2(1 + 4\mathbb{Z}) + (1 + 4\mathbb{Z}) = 3 + 4\mathbb{Z},$$
$$4(1 + 4\mathbb{Z}) = 3(1 + 4\mathbb{Z}) + (1 + 4\mathbb{Z}) = 0 + 4\mathbb{Z}.$$
Thus $1 + 4\mathbb{Z}$ is an element of order 4 in a group of order 4, and is therefore a generator of the group, so $\mathbb{Z}/4\mathbb{Z}$ is cyclic.

6.12 (a) We show that the four group axioms hold.

G1 Let $(a,b),(c,d) \in X$; then a, b, c and d are all positive.

From the definition of $*$,
$$(a,b) * (c,d) = (ac, bd)$$
and this belongs to X, since $ac > 0$ and $bd > 0$.

So axiom G1 holds.

G2 $(1,1)$ is an identity element for X; for any $(a,b) \in X$,
$$(1,1) * (a,b) = (a,b) = (a,b) * (1,1),$$
so axiom G2 holds.

G3 Let $(a,b) \in X$; then $a > 0$ and $b > 0$.
Now
$$(1/a, 1/b) * (a,b) = (1,1)$$
$$= (a,b) * (1/a, 1/b);$$
also, $(1/a, 1/b) \in X$, since $1/a > 0$ and $1/b > 0$.

So $(1/a, 1/b)$ is an inverse of (a,b).

G4 Let (a,b), (c,d) and (e,f) be any three elements of X; then
$$(a,b) * ((c,d) * (e,f)) = (a,b) * (ce, df)$$
$$= (ace, bdf) \qquad (S.5)$$
and
$$((a,b) * (c,d)) * (e,f) = (ac, bd) * (e,f)$$
$$= (ace, bdf). \qquad (S.6)$$
The two expressions (S.5) and (S.6) are equal, so axiom G4 holds.

Hence $(X, *)$ satisfies the four group axioms and so is a group.

It remains to show that $(X, *)$ is Abelian.
We have
$$(a,b) * (c,d) = (ac, bd)$$
and
$$(c,d) * (a,b) = (ca, db).$$
The right-hand sides of these equations are equal, since multiplication of real numbers is commutative, so $(X, *)$ is Abelian.

(b) To show that $A = \{(a,a) \in \mathbb{R}^2 : a > 0\}$ is a subgroup of X, we show that the three subgroup properties hold.

SG1 Let (a,a) and (b,b) be any two elements of A; then $a > 0$ and $b > 0$.
Now
$$(a,a) * (b,b) = (ab, ab),$$
which belongs to A, since $ab > 0$.
So property SG1 holds.

SG2 The identity element $(1,1)$ belongs to A.
So property SG2 holds.

SG3 The inverse of $(a,a) \in A$ is $(1/a, 1/a)$, which belongs to A, since $1/a > 0$.
So property SG3 holds.

Hence A satisfies the three subgroup properties and so is a subgroup of X.

The group X is Abelian, so all its subgroups are normal subgroups; in particular, A is normal in X.

(c) The coset $(1,2)A$ is the subset of X obtained by combining each element of A with $(1,2)$:

$$(1,2)A = \{(1,2) * (a,a) : a > 0\}$$
$$= \{(a, 2a) : a > 0\}.$$

This is the set of points in X which lie on the line $y = 2x$. Since X is the set of points in the first quadrant of $\mathbb{R}^2$, excluding the axes, we obtain the following sketch of $(1,2)A$.

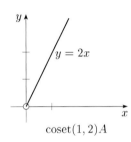

coset $(1,2)A$

(d) $(\frac{1}{2}, \frac{1}{3})A = \{(\frac{1}{2}, \frac{1}{3}) * (a,a) : a > 0\}$
$= \{(\frac{1}{2}a, \frac{1}{3}a) : a > 0\}$
$= \{(x, \frac{2}{3}x) : x > 0\}$

This is the set of points in X which lie on the line $y = \frac{2}{3}x$.

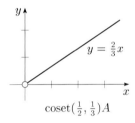

coset $(\frac{1}{2}, \frac{1}{3})A$

$(3,5)A = \{(3,5) * (a,a) : a > 0\}$
$= \{(3a, 5a) : a > 0\}$
$= \{(x, \frac{5}{3}x) : x > 0\}$

This is the set of points in X which lie on the line $y = \frac{5}{3}x$.

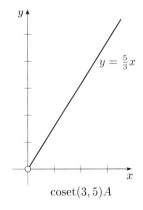

coset $(3,5)A$

(e) $(1,b)A$ is a coset of A in X for $b > 0$. We have to show that *all* cosets of A in X have this form.

We know that any coset of $A \in X$ can be written in the form $(c,d)A$ for some $(c,d) \in X$. Here $c > 0$, which implies $1/c > 0$, so

$$(1/c, 1/c) \in A.$$

Thus one element of $(c,d)A$ is

$$(c,d) * (1/c, 1/c) = (1, d/c).$$

This element of $(c,d)A$ has the form $(1, b)$ (where $b > 0$). Hence

$$(c,d)A = (1,b)A$$

and the result follows.

(f) From the formula for $\cdot$, we obtain

$$(1,x)A \cdot (1,y)A = ((1,x) * (1,y))A$$
$$= (1, xy)A.$$

(g) The group X/A consists of the set of cosets

$$\{(1,x)A : x > 0\},$$

together with the binary operation $\cdot$ defined by

$$(1,x)A \cdot (1,y)A = (1, xy)A.$$

There is a resemblance here to the group $(\mathbb{R}^+, \times)$ of positive real numbers under multiplication—we multiply together the positive real numbers which occur in the *second* coordinate of the ordered pairs. We can express this precisely, as follows.

Let ϕ be the function

$$\phi : X/A \longrightarrow \mathbb{R}^+$$
$$(1,x)A \longmapsto x.$$

Then ϕ is one-one and onto. Moreover, if $(1,x)A$ and $(1,y)A$ are two elements of X/A, then

$$\phi((1,x)A \cdot (1,y)A) = \phi((1,xy)A)$$
$$= xy$$
$$= \phi((1,x)A)\,\phi((1,y)A).$$

Thus ϕ is an isomorphism and $X/A \cong \mathbb{R}^+$.

Index